Easy Guide to
Indigenous Shrubs

*Supported by the
National Botanical Institute*

Easy Guide to Indigenous Shrubs

Pitta Joffe

BRIZA

Published by

BRIZA PUBLICATIONS

CK 90/11690/23

PO Box 56569
Arcadia 0007
Pretoria
South Africa

First edition, first impression, 2003

Copyright © in text and photographs: Pitta Joffe
Copyright © in published edition: Briza Publications

All rights reserved. No part of this publication may be reproduced or transmitted in any form or by any means without written permission of the copyright holders.

ISBN 1 875093 40 0

Disclaimer

Although care has been taken to be as accurate as possible, neither the author nor the publisher makes any expressed or implied representation as to the accuracy of the information contained in this book and cannot be held legally responsible or accept liability for any errors or omissions. This book contains numerous quoted examples of medicinal or other uses of plants. The publisher and author do not assume responsibility for any sickness, death or other harmful effects resulting from eating or using any plant in this book. Readers are strongly advised to consult professionals, and any experimentation or prolonged usage is done entirely at your own risk.

Managing editor: Reneé Ferreira
Copy-editor: Frances Perryer
Cover design: Sally Whines, The Departure Lounge
Design and layout: Alicia Arntzen, The Purple Turtle Publishing Services
Reproduction: Castle Graphics, Johannesburg
Printed and bound by Tien Wah Press (Pte.) Ltd, Singapore

Contents

Introduction 6

What are shrubs and why do people like growing them? 8

Tips on planning a garden 10

How to plant and care for shrubs 14

    Planting shrubs 14

    Feeding shrubs 16

    Keep your soil alive too! 17

    Make the most of mulch 17

    Pruning 19

Birds and butterflies bring your garden to life 20

    Birds 20

    Butterflies and moths 22

It's easy to propagate your own plants 24

    Propagating plants from seeds 24

    Propagating plants from cuttings 26

    'Potting on' seedlings and cuttings 27

    Propagation glossary 28

How to use this book 30

Small shrubs 32

Medium shrubs 68

Large shrubs 93

References 126

Index 127

Introduction

In my previous two books I emphasised the beauty, variety and usefulness of our indigenous plants: trees, shrubs, climbers, perennials, annuals and water plants. Since then, so many people have expressed a need for something concentrating on our lovely South African shrubs that I have adapted them to make a new, pruned-down book with more specific information on shrub cultivation, introducing 12 species previously omitted.

The idea behind this book is that it will provide a good selection of indigenous shrubs for people who are starting their 'gardening lives'. I have chosen plants that are, on the whole, easy to look after and obtain. About 12 are distinct challenges, because challenges are something that every gardener has to face at some point or another. If you live where these plants would thrive, try some of them. Don't be afraid to experiment – if you're a beginner, start with some of the really easy ones, particularly those that tend to grow naturally in your area. As you build up confidence with your successes, so you can move on to more challenging plants. Most of the shrubs in this book are easy to grow; detailed instructions have been given to help you.

Fairly comprehensive instructions have also been provided on how to propagate some of the plants, either from seed or from cuttings. Use them to gain experience before you start experimenting on some of the other plants. This is not only enormous fun, but also provides the deep satisfaction of seeing your tiny seeds germinating, and later thriving as 'adult' shrubs in the garden. It is also an excellent and rewarding hobby for people who have time on their hands and would like to grow something new to give away to friends. Young homeowners struggling to make ends meet might like the challenge of propagating, even selling, their own shrubs, while people living in remote areas might want to grow indigenous plants from seed collected from the surrounding bush. This is essentially a starter book. Once you have gained a little experience cultivating some shrubs, you can start experimenting with others.

This book is perfect for people with small to medium-sized gardens. It also caters for people with large and farm-sized gardens. In smaller

LEFT: Irresistible! South Africa's Impala Lily (*Adenium multiflorum*).

(e.g. townhouse) gardens, you have to remember to scale down your thinking – you can't plant a 30–40 m tree there. Your motto should be to think smart and scale down! In a small garden, your large shrubs can become your 'trees', your medium shrubs become your 'large' shrubs, and your small shrubs can be regarded as 'medium' shrubs. If you have chosen carefully you probably won't need too much else – but if you prefer lots of annual or perennial colour (e.g. *Dimorphotheca* or *Felicia*) there are hundreds of plants to choose from at nurseries.

In this way you can garden without any large conventional trees in a small garden. A number of the large shrubs eventually mature into small trees if conditions are suitable. Some of the fast-growing species are bushy at first but may become leggy (i.e. bare at the base) as they mature. These plants must either be pruned regularly to prevent this, or can be encouraged to develop into small trees perfectly suited to the size of the garden. Large shrubs like the False Olive (*Buddleja saligna*) can be trained into a tree shape from a young age. Otherwise they can be left to develop into shrubs, but if they are not regularly pruned back and shaped, their lower branches will become barer as they slowly mature. At some stage, a decision will have to be taken to remove the lower branches. This opens up a whole new gardening opportunity – the area under the 'tree' will need a new set of plants. I have purposely included a number of such large shrubs that can be trained into small trees for use in smaller gardens. One is not actually replacing trees as such, just scaling down one's proportions to suit those of a very small garden. I've seen some scary landscaping in pocket-sized townhouse gardens, with large trees pushing and shoving to escape, elbowing against windows and drainpipes, blocking out every bit of sun and no doubt damaging foundations and paving in their efforts to abscond!

Top: *Dimorphotheca ecklonis*.
Middle: *Dimorphotheca jucunda*.
Bottom: *Felicia amelloides*.
All perfect for providing temporary colour in a new garden.

What are shrubs and why do people like growing them?

Gardeners could hardly wish for a more versatile group of plants than indigenous shrubs. Well chosen, these plants can be used to design unique and beautiful gardens. They range in height from extremely low (30–40 cm), when they can be used as groundcovers, to about 5 m in height. Some are soft-wooded while others are fairly woody. By definition, shrubs are 'bushy' from ground level, tending to form branches low down, whereas trees generally have long, bare trunks. Many are evergreen, while others are deciduous (lose leaves in winter).

Shrubs offer such a diversity of leaf shapes and textures, foliage and flower colours that their effective use will contribute greatly to making your garden interesting and attractive. The fruits, seeds and nectar of most of these plants attract a wide variety of birds and this alone makes them

Clearly birds are enjoying these fruits of the Tulip Tree (*Thespesia acutiloba*).

worth growing. Some are treasured for their profusion of flowers, e.g. Wild Bush Petunia (*Barleria greenii*) and Cape Honeysuckle (*Tecomaria capensis*), while others are valued for their lovely foliage, e.g. Forest Bells (*Mackaya bella*) and Pistol Bush (*Duvernoia adhatodoides*), and still others for their bright berries, e.g. Star Apple (*Diospyros lycioides*) and Tulip Tree (*Thespesia acutiloba*).

When a garden is being landscaped, trees are usually used in smaller numbers than shrubs and are planted to provide some height. Shrubs are used to fill in under the trees and screen out

A small area with a wealth of shapes, textures and colours.

things like roads and other houses. But if there is a lovely view from the garden, people generally try to frame it with trees or large shrubs and plant lower shrubs in front of it so that the view is not lost. Shrubs serve a variety of functions; they can screen strong winds or block unsightly views. They can be used as hedges or planted merely for the beauty of their flowers. Some attract birds while others are planted for their lovely fragrance. The choice is entirely up to the owner of the garden.

Gardens are generally planned with mixed herbaceous borders. These can be filled with a variety of interesting plants all serving different functions. Some are grown for cutflowers, while others have a lovely fragrance at night and attract moths. Many are planted to attract birds to the garden, providing the owners with many happy hours of pleasure and satisfaction. It is important to provide different habitats and levels of foliage to encourage as many species as possible. Some birds like to live high up (sunbirds, louries); others (Cape Robin, Burchell's Coucal) enjoy scratching low down in thick shrubs, while others are somewhere between (Southern Boubou, Cape White-eye).

To simplify your choice the shrubs discussed in this book have been divided into categories of small, medium and large, according to their height. When planning the layout of your garden, always bear in mind the ultimate size of a shrub when fully grown and allow enough room for the plant to spread so that it can reach its full potential without having to be pruned merely to control its size. The approximate height and spread of each shrub have been given in the text to assist you in your planning.

Some shrubs grow rather slowly at first. This is natural – it takes time for the root system to become well established. Once established, they will grow at a more steady pace. This is why it is sometimes best to fill gaps with faster growing temporary plants which can include annuals and perennials.

Top: Foliage texture and shape is important. The large soft and hairy grey leaves of *Barleria albostellata* create quite an impact.
Right: Plant heights used effectively in a mixed herbaceous border.

Tips on planning a garden

Home gardens are generally landscaped to imitate forests or woods. The vegetation is always in three main layers – tall trees, a medium stratum of vegetation and smaller low-growing plants that provide colour and serve as groundcovers to protect the soil. In a garden, trees are generally planted to provide vertical height, shade, a focal point and to block unsightly views. One must always be aware of the vertical design of the garden as well as the horizontal. Shrubs are on a more human scale, helping to relate the height and bulk of the trees to the garden as a whole, and they soften stark fences. Trees can be regarded as the skeleton while shrubs provide body to the design. Shrubs give gardens shape and structure and provide a sense of enclosure.

If you want a permanent green backdrop to your garden, choose evergreen shrubs. This will provide year-round privacy. In front you could vary the types of groundcover (e.g. *Felicia amelloides* and *Dimorphotheca* species), annuals or bulbs year after year to give interest to the garden. Smaller deciduous shrubs can be planted in front that are perhaps prolific flowerers, or have attractive fruits or foliage. Privacy in gardens is extremely important, especially in recreational areas where you want to relax with family and friends. These areas must be shielded from the street and from the neighbours. You can decide whether you want a solid barrier or a light screen. For a solid dense screen that will block a view completely, closely plant three Blue Honeybells (*Freylinia tropica*), or use a closely clipped Dogwood (*Rhamnus*) or False Olive (*Buddleja saligna*) hedge or choose a shrub with thick foliage like the Dune Crowberry (*Rhus crenata*) or Bushtickberry (*Chrysanthemoides monilifera*). For a lighter screen or to break up a view, plant shrubs with slightly less dense foliage, like the African Dogrose (*Xylotheca kraussiana*), Curry Bush (*Hypericum revolutum*) or Crossberry (*Grewia occidentalis*).

Shrubs can be clipped or trained into hedges to provide some privacy, but remember that this imparts a formal atmosphere to the garden. Clipping increases the density of the barrier and serves to define a boundary. Clipping the growing shoots on top and on either side of the shrub encourages the plants to grow towards one another, knitting them into a continuous row. Repeated clipping results in growth to the sides and a denser growth than in untrimmed plants. If you do plant a formal hedge, leave enough room (about 30–50 cm) between it and other plantings to allow access for pruning purposes. Evergreens make the best hedges. If privacy in winter is not the main concern, deciduous shrubs that flower in autumn or have pretty autumn colours may be chosen instead.

A closely clipped Dogwood (*Rhamnus prinoides*) hedge used to accentuate a pathway.

Large-leaved shrubs used effectively to create an impact in a largish tropical garden – Wild Date Palm (*Phoenix reclinata*) and Natal Wild Banana (*Strelitzia nicolai*).

Evergreen or deciduous may be used to advantage in other ways. Think carefully where you would like a little sun in winter. In these areas one can easily plant a deciduous shrub like the Parasol Flower (*Karomia speciosa*). Sometimes a little extra sun and warmth are worth a loss of privacy in winter. Use a fast-growing evergreen like Karoo Sage (*Buddleja glomerata*) to screen the view into a bedroom window – don't plant it right next to the house. It could give problems being too close to the window, like blocking access to drain pipes and windows for cleaning, etc.

When planning a garden never ignore the width of a plant! It may, for a while look charming, but at some stage one of these plants will take over and smother all the others around it.

Foliage texture and shape are important. The visual impact of tiny leathery grey leaves is very different to that of large dark-green glossy leaves. Small-leafed shrubs are generally better suited to smaller gardens, while large-leafed shrubs can be used for large expanses of shrubbery in a bigger garden.

Foliage colour also plays an important role in garden planning. Colour should be used carefully, in bursts, and at points where the coloured leaves draw attention to some feature or display. Use them to define different areas in the garden, but make it a subtle change, not a jarring one.

Aromatic foliage is a third and very lovely dimension. Shrubs with aromatic foliage are popular choices for the garden. Their impact is not only visual: they add elements of smell, taste and touch. Some can even be used to flavour meat like Wild Rosemary (*Eriocephalus africanus*) and Brown Salvia (*Salvia africana-lutea*). Others like the Rose-scented Pelargonium (*Pelargonium graveolens*) are used to flavour cakes and puddings.

When planning your garden, don't ignore the width of the plant, or you will end up with a terrible mess that you won't easily sort out or control. The plants will push and shove one another until not one of them looks like anything at all – no natural shape or form! Allow enough room for shrubs to spread freely and show themselves off beautifully. So, resist the temptation to plant too closely. Use temporary short-lived plants to fill the gaps: there are many in this book, like Golden Daisy (*Euryops pectinatus*) and Honey Euryops (*E. virgineus*), White Bristle Bush (*Metalasia muricata*), Cancer Bush (*Sutherlandia frutescens*), Purple Broom (*Polygala virgata*) and Eight Day Healing Bush (*Lobostemon fruticosus*).

The lovely honey-scented flowers of the Honey Euryops (*Euryops virgineus*) attract masses of pollinators.

Plan your garden – don't just let it happen!

Because we are individuals, gardens are unique reflections of our own personalities and likes and dislikes. If they are to do everything for us that we want from a garden, careful planning will be needed. It is a good idea to draw a careful plan to scale on paper before getting down to work. Use circles to denote width of plants to ensure they are correctly spaced. The design must take people into account as well as the plants. Let pathways follow the route that people naturally walk; don't try to force them to follow complicated zig-zags. The smaller the yard, the simpler the plan should be. Don't use hundreds of different plants in a garden, it may make the design too busy. Rather choose a smaller selection of shrubs and use larger numbers of the same species. Don't dot them all over the place, keep them together in groups or stretches, allowing them to flow around corners carrying the eye with them. If dotted all over they make the eye jump from one to the next, giving a choppy effect, and they lose their impact as a result. Don't make your plan too complicated; remember that 'less is more'. The simpler the design, the more effective it will be. Neatly clipped hedges give the garden a 'formal' feel; untrimmed shrubs, large and small, give the garden a 'wilder' feel.

Lowveld Bauhinia (*Bauhinia galpinii*) planted as a decorative screen.

WHAT CAN WE USE SHRUBS FOR?

- as fillers in the garden
- to soften hard structures (brick and prefabricated concrete walls, buildings)
- to help screen out unsightly views
- to block the view of the neighbours
- as screens against the wind
- to screen the pool from the rest of the garden
- as hedges to define boundaries
- as groundcovers (especially small shrubs)
- as formal hedges for screening and for privacy (low shrubs planted in front of taller ones give privacy from the ground up)
- as low hedges or curved hedges to enclose a sitting area
- to provide homes and food for birds and butterflies
- many have fragrant flowers or foliage
- they make lovely cutflowers
- to provide fruits for eating and making jams and preserves
- many have medicinal uses
- to create a lovely garden to look at, sit in and enjoy!

A garden designed by Nature! Front left: *Tripteris oppositifolia*. Middle: Blou Fluitjiesbos (*Lebeckia sericea*).

How to plant and care for shrubs

Planting shrubs

Shrubs can be planted directly into well-prepared holes, but usually grow more vigorously if planted into well-prepared flowerbeds which have been enriched with plenty of compost and organic material. Transplant young shrubs carefully to minimise transplantation shock and to ensure that the roots become established as soon as possible. Give your plant a good start in life by preparing the hole well.

TO PLANT IN WELL-PREPARED HOLES

Dig a large square hole at least twice the width and depth of the nursery bag (about 60 cm × 60 cm × 60 cm or 1 m × 1 m × 1 m in poor soil), keeping the topsoil separate from the subsoil. Loosen the soil in the bottom and corners of the hole and water well. Do this a couple of times, saturating the soil well. Add the following to the topsoil:

- compost, equivalent to about half or more the amount of topsoil removed
- 280 g (250 ml or 1 cup) of superphosphate and/or 1 cup of bonemeal
- 150 g (250 ml or 1 cup) of slow-release 3:2:1 fertiliser.

Mix thoroughly. Place approximately one third of this mixture into the hole. At this stage place the plant, bag and all, in the hole to test whether the soil levels in the bag and on the ground are the same. Remove the plant and water the hole again. It is easier to saturate the soil in the hole at this stage than it is after you have planted the shrub.

Remove the shrub from its container, taking care not to damage the roots. Only if the root-ball has become very compacted should it be loosened. Hold the shrub upright (vertical) in the hole, even if you have to tilt the root-ball to achieve this, and add the remaining topsoil mix. Tamp the soil gently around the root-ball, taking care not to compact it. If necessary, add compost to the remaining subsoil and use to fill the hole. The soil below and around the root-ball should be firm enough not to settle excessively after watering. Ensure that the top of the root-ball soil aligns with the surrounding soil once the shrub has settled. Note: Superphosphate and bonemeal are 'root foods', ensuring strong, healthy root growth; 3:2:1 has a high proportion of nitrogen for lush green leaves and fast shoot growth (see page 16).

Make a generous basin around the trunk, at least as wide as the shrub's aerial spread and 15 cm deep. Mulch with a 10 cm thick layer of organic material. Water thoroughly directly after planting by allowing a hosepipe to trickle gently into the hole; and water carefully for the next four to six months, especially in dry weather. Thoroughly saturating the hole in this way should give the young shrub a very good start in life and it shouldn't need water for quite a few of weeks. It will send its roots downwards after the moisture below. Good, deep watering (at least 10–20 litres at a time) is essential. Shallow watering (wetting only the top 5–10 cm of soil) encourages roots to remain near the surface of the soil instead of growing down deeply to find the moisture below. A plant with a deep, well-developed root system tends to be well-anchored against strong wind and to have some drought resistance.

TO PLANT IN A WELL-PREPARED FLOWERBED

Shrubs usually grow more vigorously if planted into well-prepared flowerbeds. To ensure healthy plant and root growth, thorough preparation of the soil prior to planting is essential. It is particularly important to add as much organic material as possible to the soil – you cannot add too much! In loose, sandy soil the organic material helps to retain water, and in heavy clay soil it serves to aerate the soil and improve drainage and water absorption.

Having marked the outlines of a proposed flowerbed (use a flexible sun-warmed hosepipe and make generous sweeping curves), dig the bed over to a depth of about 40–50 cm, breaking up all large clods of soil. Spread a thick layer of organic material and compost (about 30 dm^3, or one bag, per square metre) evenly over the surface, and sprinkle approximately 200–250 ml each of slow-release 3:2:1 fertiliser, superphosphate and/or bonemeal per square metre, on top of the compost. Dig the whole bed over again to incorporate the compost and fertiliser. Saturate the bed using a fine or gentle sprinkler. Allow to settle for a few days to a week. Plant directly into the bed.

Water thoroughly directly after planting, using a fine sprinkler, and thereafter two to three times a week in very hot, dry and/or windy weather and once a week in cooler weather. Well-established trees and shrubs should only need water about once a month in winter. In summer, water in the early morning or evening – less water will evaporate in the sun. In winter, especially in cold frosty areas, water in mid-morning so that the leaves can dry before the evening otherwise the water may freeze and the plant may die.

A good, deep watering is more beneficial than numerous short shallow sprinklings. If a bed contains only shallow-rooted plants such as annuals and groundcovers, saturating the top 10 cm of the soil should suffice, but a bed which also contains shrubs and trees needs much deeper watering. Allow the top 5–7 cm of soil to dry out before watering again. Constantly oversaturated soil prevents air from reaching the roots and could lead to root rot, killing the plants. It is important to remember that plants originating in the winter rainfall region have to be watered through the winter if grown in the summer rainfall area and vice versa.

Plants should be fertilised at intervals of 6–8 weeks throughout the growing season, using slow-release 3:2:1 for foliage plants and 3:1:5 for flowering plants (see page 16). Bulbs should be fed from the time that they stop flowering until they become dormant. This is when they build up reserves for the next flowering season. Always mulch flowerbeds well with a 10 cm thick layer of organic material and replenish regularly.

A selection of organic and inorganic fertilisers.

Safe ways of dealing with pests

Smaller pests that may prove to be a problem are aphids, white fly, scale insects, Australian bug, thrips, mealy bugs, red spider mite and ants. Thrips (blackish flying insects) and spider mite are very small – almost invisible to the eye. White fly (tiny white flies), aphids (soft, roundish sucking insects) and mealy bugs (look like small specks of cottonwool) are slightly larger and can be seen fairly easily if you examine the plant carefully. Australian bug (a soft scale insect with an oval, fluted powdery white back) and scale insects (brown or grey, hard-shelled, round or oval) don't move and are somewhat smaller than a pea. Most of these insects suck sap, causing foliage and flowerbuds to become distorted, discoloured, curled or twisted. Growth of the plant may be stunted and it may even die if the infestation is heavy enough.

A good safe spray that is safe for children, pets and wildlife can be made by mixing one teaspoon of bicarbonate of soda and one teaspoon of washing-up liquid ('Sunlight') in a litre of water. Do not use detergent. Shake well, and put it into a bottle with a spray nozzle (e.g. an old 'Windolene' bottle that has been thoroughly washed). Spray the mixture onto pests – it will kill ants and aphids. Don't spray on the 'good' insects that help to rid the garden of pests! After spraying, feed the plant – healthy plants resist pests and diseases!

Aphids are common pests and can do damage sucking sap and spreading plant disease. Spray them with the safe spray. Australian bug and scale insects can either be picked off by hand or sprayed with a toothbrush dipped in the safe spray. The Australian bug is easily squashed because it has a soft body. Dip an earbud into methylated spirits and put that onto the Australian bug or scale insect – don't get the meths onto your plants. Spider mite make a silvery webbed pattern on the leaves of the plant. Spray them with the safe spray every day. Remove the plant to a moister, cooler area. Remove mealy bugs as soon as they appear – use cottonwool dipped in meths.

Mealy bugs, aphids, white fly and scale insects make leaves mottled and sticky with honeydew, which attracts ants and causes further problems. The ants love this honeydew and 'farm' certain of the insects for it. They move insects from plant to plant ensuring that the insects are in prime positions. In this way the infestation is spread from one plant to another and you have a big problem. If you see ants running up and down your plants – know that you have a problem of one sort or another. Deal with the ants quickly by spraying with the safe spray and move the plants out of the way of the ants if possible. Another way to deal with ants is to mix together equal quantities of castor sugar and borax. Sprinkle this mixture near ant nests and around the house. Sugar attracts the ants and borax acts as an abrasive, destroying the ants' dehydration protection.

Snails can be caught and squashed or you can sprinkle salt onto them to kill them. Spread a layer of ash around your plants to discourage snails and slugs. You could also sink a shallow dish or cup, filled with beer, into the soil. Snails and slugs are attracted by the smell – they fall in and they drown because they can't get out.

One could also mix together 1 teaspoon dried yeast, 1 cup warm water and 1 tablespoon sugar, place in a foil 'pie dish' and place in the garden as above to achieve the same result.

Feeding plants

Feed with slow-release fertiliser (3:2:1 for foliage shrubs and 3:1:5 for flowering shrubs) at intervals of 6–8 weeks throughout the growing season. Do not feed in winter – this encourages soft young growth that is very susceptible to frost damage. Shrubs situated in a lawn need to be fed and watered particularly well as the grass will compete with the plants for moisture and nutrients. Leave a grass-free circle around the base of the plant to facilitate mulching, watering and feeding.

Foolproof fertilising

The nutrients in humus, compost and old kraal manure may not be sufficient for normal plant growth. Suitable fertilisers should then be used, strictly according to the instructions on the label. Fertiliser formulas are easy to understand once you know the secret! A vast array of organic and inorganic (chemical) fertilisers is available at nurseries and hardware shops. Most have numerical formulas that look daunting, but are really simple to understand once you know what each number represents. The numbers indicate the proportions (ratios) of each of the three main plant foods contained in a particular fertiliser. The formula is always written in the same order: the first number always represents the proportion of nitrogen; the second number, the proportion of phosphate; and the third number, the proportion of potash. An example of a high-nitrogen fertiliser follows:

3 : 2 : 1

3 parts nitrogen – foliage food, speeds up growth and results in lush, deep-green leaves.

2 parts phosphate (superphosphate) – root food, results in strong root growth (dig bonemeal and/or superphosphate into the ground where roots can reach it, as it is not too soluble).

1 part potash (potassium) – flower/fruit food, results in more flowers/fruits, increases disease resistance and improves the quality of plants or crops.

Plants need all three (plus trace elements) in approximately the right proportions. If one is missing, overly high doses of the others will not compensate, and the plants' growth will still be impeded. So fertilisers must be bought in the right proportions to suit a particular soil's needs – and soils differ from area to area. For example, some soils are deficient in nitrogen and will need a high-nitrogen fertiliser to compensate. Acid soils are often low in phosphates and will benefit from an application of superphosphate. Soils that are too alkaline or acid may prevent nutrients in the fertiliser being actively absorbed by plants. Experiment with the different ratio fertilisers until you find the one that works best, or ask the local hardware store or nursery for advice. Your plants will soon show which formula is most suitable by growing fast and flowering well. Allow at least 6 weeks for some reaction.

Keep your soil alive too!

Leaf litter or humus is composed of all the leaves, twigs, dried flowers and fruit that drop from plants during the natural course of growth. They decompose where they fall. This is exactly what nature intended. In a healthy and thriving forest there is always a thick litter layer that both protects the soil and feeds the plants. Take a careful look at a natural indigenous forest floor next time you hike through, and pick up a tip or two for your own garden. If you want your garden to look as healthy as that forest, provide it with an equally thick layer of mulch. You can either leave plant material where it falls, or add chopped-up prunings to the layer to increase its depth. Bark chips, mulch or compost can be purchased and used as a slightly neater option. In a healthy, thriving garden the soil is covered with a thick layer of decomposing mulch that smells rich and healthy when run through the fingers.

Encourage millipedes, they play a vital role in the decomposition process.

Encourage a healthy population of creatures in this mulch layer. They provide a vital service in the decomposition process, especially earthworms, millipedes and beetle larvae, and at the same time become food for birds (thrushes, robins) that scratch about in the leaf litter. Thrushes will pinch some of the mulch as nesting material! It

Certain termites and beetle larvae are responsible for the decomposition of wood, releasing nutrients back into the system.

is mainly beetle larvae and termites that break down leaf litter into units small enough to be finally broken down by bacteria and fungi, releasing or recycling the mineral nutrients back into the soil. Certain types of termite and beetle larvae are also responsible for the decomposition of wood (fallen trees), again releasing nutrients back into the system. Woodpeckers and barbets will spend hours tapping and probing into dead branches looking for these insects, a real treat! However, do take care that insects such as termites don't become pests inside your home. Keep insects outside where they belong!

Make the most of mulch

Bare soil is not only unsightly, it is also extremely unhealthy as far as your soil is concerned. It is equivalent to you spending your life exposed to the elements without sunscreen or any other protection. The harsh sun leaches nutrients from the soil and the impact of water droplets (from rain, hosepipes or sprinklers) compacts the surface until it is about as porous as concrete! Concrete doesn't absorb water – and nor will your flowerbed, in this condition. This, of course, leads directly to another very wasteful (you or your gardener's time) and unhealthy (for your garden) practice, that of 'digging over' beds – a pastime that dates back to the Dark Ages! The freshly dug soil is now just waiting to be washed away by the next rainstorm or windstorm! South Africa loses billions of tons of topsoil every year due to practices such as these. Help to conserve our soil and its organisms by gardening in a 'nature-friendly' and caring way. Do your bit towards conservation by protecting your own soil and the incredibly rich and diverse web of life it supports.

Mulch acts like a protective 'blanket' reducing evaporation and keeping the soil surface and roots cool and moist.

Mulch is nature's way of feeding plants, conserving moisture and protecting the soil surface and structure. It acts like a protective 'blanket' reducing evaporation and keeping both the soil surface and roots cool and moist. The opposite is true in the cold winter months, when soil temperature is kept at a reasonable level, protecting roots from frost damage. Mulch prevents soil from being carried away, either by wind or rain. It 'cushions' the sometimes considerable impact of water droplets (avoid watering with a fully open hosepipe or bucket), keeping the soil surface soft, crumbly and porous. Both water and oxygen are then able to penetrate the soil easily, thereby reducing water usage. Plant roots need to 'breathe' in order to thrive. The tunnelling activities of earthworms, as they fetch and carry organic material, help to aerate the soil and their droppings enrich it. This, together with the activities of soil microorganisms, produces a soft, rich and crumbly soil that roots will love! Lastly, mulch helps to some extent to suppress the germination of weeds.

Don't sweep up and then proceed to throw away fallen leaves, dried and faded flowers, fruits, small twigs, etc. Apart from the fact that this is a totally misdirected use of energy, it is also equivalent to throwing away bags of compost – something I'm sure no one would ever contemplate doing! A far healthier scenario, for you and your plants, would see that same energy being channelled into spreading the organic material over the surface of the soil in your flowerbeds, or under trees and shrubs – wherever bare soil is visible. Ensure that the layer is nice and thick (at least 5–10 cm). Feel free to use lawn 'shavings' or clippings from the lawnmower box – a thin layer, not including roots and runners, any disease-free, chopped-up prunings or seed-free weeds to top up the layer. Other suitable organic materials include dried leaves, bark chips under trees and shrubs, hay, straw, coarse compost, chopped up twigs and branches which can be hidden under shrubs with dense foliage in the 'bird garden', vegetable peels and fruit skins which you can hide under shrubs with thick, ground-level foliage, well-rotted manure, pine needles for acid-loving plants such as proteas and ericas, sawdust and bought mulch. Replenish the layer regularly, because it decomposes with time, providing your plants with lots of nutrients. Remember too, that it feeds the soil fauna that recycle the nutrients back into the soil – a soil in excellent condition is 'crawling' with earthworms and other organisms. When spreading the layer, keep it about 5–6 cm away from the stems of shrubs and tree trunks to prevent fungal diseases.

Before applying mulch for the first time, especially if the soil is in a 'concrete' condition, dig the bed over lightly. Apply a thin (4–5 cm) mulch and compost layer and a sprinkling of slow-release 3:2:1 fertiliser, superphosphate and/or bonemeal. Using a garden fork, dig the top layer over well (about 30 cm deep), to incorporate as much of the organic material as possible. Now add a second slightly thicker mulch layer to cover and protect the first one. Dig in lightly with a fork. From now on it will only be necessary to top up the mulch layer regularly. When preparing a new flowerbed, dig as much compost and other organic material as possible into the top 20–30 cm of soil. Before digging over, sprinkle bonemeal (and/or superphosphate) and slow-release 3:2:1 fertiliser at the rate of 1 cup (250 ml) per square metre. Water the bed well using a gentle spray, allow it to settle for a week or so, and then plant your trees and shrubs directly into it. After planting, apply a thick mulch layer to protect the well-prepared soil, and irrigate young plants thoroughly, again with a very gentle spray. All that remains is to top up the mulch layer regularly, and to watch your garden grow well and lushly. If unconvinced, experiment somewhere in your garden with this natural method of gardening – you'll soon be converted!

Pruning

Pruning can be used both to extend a shrub's lifespan and to ensure the regular formation of better quality flowers. One also prunes to remove old, dead or diseased wood. Some shrubs require only occasional trimming to remove dead flowers or branches to keep them neat. Resist the temptation to prune too much and too vigorously. It is usually better to prune after flowering. Use good quality, sharp secateurs and make the cut just above the leaf bud. Choose a bud pointing in the same direction as you would like the new branch to grow, usually upwards and outwards. Make the cut about 5 mm above the bud and cut at an angle of 45° down and away from it (see illustration).

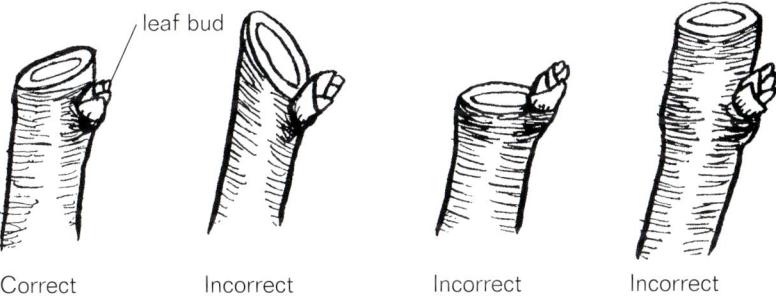

Correct Incorrect Incorrect Incorrect

Frost-damaged plants should not be pruned until spring has well and truly arrived – the damaged, frost-burnt foliage protects tender growing points. When removing an unwanted branch, cut flush against the healthy trunk without leaving a stump. This not only looks ugly, it may also die back. Make a clean cut, first saw upwards from underneath and then downwards, from above to meet the bottom cut. This will prevent a strip of bark from being torn off when the branch falls. It is wiser to remove unwanted shoots while young, with secateurs, rather than allowing them to develop into large, unmanageable branches. Large, overgrown shrubs that have not been pruned for years should be cut back gradually, bit by bit, to ensure that over-enthusiastic pruning does not kill them.

How *not* to prune!

Birds and butterflies bring your garden to life!

Symbols are used throughout your book to indicate good 'bird' and 'insect' plants and to help you spot them at a glance. Read the information given for each species – this will help you to make the best choices for your garden. Consult, research or observe to discover what plants are best suited to your own area. Enjoy your birdwatching!

Birds

Careful planning and a little work could soon have you listening to a variety of wonderful bird songs. A selection of the right indigenous food plants ensures that your favourite birds will visit or move in (expect birds that occur naturally in your area). Sunbirds, for example, are unlikely to visit unless you have nectar plants. Waterbirds won't stream in without a nicely situated pond or dam. Large well-placed flattish rocks, some fish, frogs, tadpoles and a bit of riverine vegetation will make the place irresistible. Fruit and insect eaters won't appear unless trees and shrubs produce fruits and flowers that attract insects – otherwise why visit at all?

Scarletchested Sunbird feasting on Cape Honeysuckle (*Tecomaria capensis*) flowers.

Birds spend a lot of time searching for food, especially when they have chicks to cater for. They tend to live where their needs are easiest met. If nothing in your garden tempts them to stay, they will merely pass through on their way elsewhere. The secret to a successful bird garden is to plant a mix of local and other plants – ones that you know will do well in your area. Accordingly, they will grow fast and well, they will meet the needs of the local birds and will look good in your garden. They will be strong and healthy and relatively trouble free. Our local plants also support higher populations of insects (bird-food) than do the exotics.

Entice birds to your garden by providing suitable habitats, safe nesting and roosting sites and their own natural food – grasses, trees and shrubs that they normally make use of. Grow plants whose flowers and nectar attract hordes of insects – most urban birds will feed their young on insects. The

Whitebellied Sunbird sipping nectar from an erythrina.

Blackheaded Oriole and Crested Barbet will stop by for a sip of nectar if you have planted a Mountain Aloe (*Aloe marlothii*). The bright-red flowers of the Dwarf Coral Tree (*Erythrina humeana*) are a rich source of nectar and attract Blackheaded Orioles and Whitebellied and Black Sunbirds. Nectar-rich plants such as aloes attract not only brilliant sunbirds, but also the Crested Barbet, Cape White-eye, Blackheaded Oriole, Blackeyed Bulbul, Streaky-headed Canary and Grey Lourie. These birds all visit the Krantz Aloe (*Aloe arborescens*), and the bulbuls, louries and canaries eat the flowers! Aloes not only provide nectar, but also striking colour in a dull and tired-looking winter garden. In autumn, sunbirds sip nectar from the Cape Honeysuckle (*Tecomaria capensis*) and Wild Dagga (*Leonotis leonurus*). These decorative, easy-growing shrubs are a beautiful addition to any garden and tide nectar-loving birds over until the aloes start flowering in winter.

Birds can't resist the Tree Fuchsia (*Halleria lucida*), which often flowers and fruits at the same time. It is a perpetual source of food for them. Many love it for its nectar, others eat the flowers or fruits or the insects attracted by the flowers. Frequent visitors include Crested and Blackcollared Barbets, sunbirds, Cape Robin, Cape White-eye, Blackeyed Bulbul and Burchell's Coucal, which often hunts for treats on the ground at the base of this thick shrub.

Cat's Whiskers (*Clerodendrum glabrum*) produces small flowers that attract a variety of insects, especially bees and butterflies. Each year many of these butterflies become food for birds. Apart from being 'bait', butterflies are beautiful to have in a garden and are excellent pollinators too, making it possible for your flowers to produce fruits and set seed. It is better not to use insecticides in your 'bird-friendly' garden – you might end up poisoning the very birds that are helping you to keep it pest free. Pests are their natural food and each of them consumes hundreds every day. Remember, you cannot have butterflies without caterpillars – feeding caterpillars is a small sacrifice to make for the privilege of having brilliant butterflies in the garden. It is better not to spray the caterpillars – they only last a short while and provide food for a host of interesting birds, such as the migrant cuckoos who travel all the way from Europe especially for this treat. The munching of the caterpillars can be regarded as a form of natural pruning – the plant will soon sprout a brand new set of leaves in fresh, spring green, like our Canthium.

Some birds like thickets and dense bush to nest and hunt in, while others stick to the treetops. Each bird is most at home at a particular height, although some birds inhabit various levels. Forest birds are usually only comfortable in the tops of thick forest trees. Try to copy nature in different parts of the garden. If you live where forest trees and birds will thrive, try planting a small patch of forest. You can even use some of the local forest species – remember to 'layer' your forest – true forest has layers of vegetation at different heights. Or plant a small patch of 'riverine forest' using taller and shorter trees to form layers at different heights. Use shrubs, perennials and small multi-stemmed trees or large shrubs such as the Wild Honeysuckle Tree (*Turraea floribunda*) to form the shorter or lower layers between the taller trees. Do not plant rows of tall trees that will all eventually reach the same height – few

Brownheaded Parrot.

birds enjoy that. The more layers you create the more birds you are likely to have. Some birds, like the Cape Robin and Burchell's Coucal will creep about in lower layers – they like dense thickets near the ground. Any plant can be used for this purpose as long as it forms thick bush and is not too tall. The Wild Honeysuckle Tree (*Turraea floribunda*) is well suited – its prolific flowers smell rather like gardenias and the wonderful scent permeates the garden. The attractive, bright orange-red seeds are incredibly popular with birds and disappear in no time at all.

Don't pick up the leaves, faded flowers, twigs, etc. (this, in fact, goes for the rest of your garden too!) that drop to the floor of your 'forest'. Leave it exactly the way it is, the way nature intended. A mulch layer protects the soil and plant roots, and also decomposes into compost to provide nutrients for the plants. The real bonus is that all sorts of creepy-crawlies live and hide in there – in a word, potential bird-food! Don't disturb the area – allow insectivorous birds to search about in there for a treat. Don't trim away the lower branches of the shrubs and perennials – they shelter shy birds. One or two decomposing logs will house beetles, spiders, earthworms and other small creatures – all prospective food. A pond or large shallow birdbath (but only 1–2 cm deep) will do a lot to encourage even more birds.

Butterflies and moths

South Africa has some really magnificent butterflies, and apart from the fact that they are important pollinators enabling flowers to produce fruit and set seed, their beauty alone should persuade us to encourage them to our gardens. Apart from sipping the occasional bit of nectar for energy and pollinating flowers as they go, butterflies have a very important function in life: to mate and for the female to lay her eggs on specific plants called host plants that are suitable for 'her' caterpillars to feed on. Each butterfly species has host plants that are specific to it. After mating, the female must begin the hunt for suitable plants, and will move from area to area or garden to garden, gaining strength by sipping the odd bit of nectar from particular flowers along the way. She will continue her journey until she has located a suitable host plant in good enough condition to support hungry caterpillars.

The African Monarch (*Danaus chrysippus*) carefully tests leaves for freshness using sensory cells in the forelegs and antennae. The first thing the young caterpillars do after hatching is to eat their egg-cases for a protein boost, but after that they tuck in to their favourite host plants.

The fact that most butterflies are host-specific makes them very vulnerable. If we are not careful to protect and conserve our heritage of indigenous plants it could lead to the extinction of many of our wonderful butterflies. Butterflies not only require specific plants on which to lay

Diadem butterfly (female) visiting Plakkie (*Crassula ovata*) flowers.

The caterpillars of the Specious Tiger Moth feed on *Ficus* and *Carissa* species.

their eggs, but also develop a preference for plants in a particular area and will visit them year after year. In the wild, the estimated mortality rate of eggs and larvae is 98%. Pesticides and pollution are also a constant threat. Make your garden a haven for butterflies by providing both host and nectar plants, and lots of sunshine to warm them!

Cultivate a wide selection of host plants and a variety of beautiful free-flowering shrubs and perennials to supply nectar refreshment. Caterpillars rapidly grow into butterflies and the plant will soon boast a brand new set of bright and shiny leaves. Some plants seem to be favourites for butterflies. Examples are *Clerodendrum glabrum*, *Crassula ovata*, *Bauhinia galpinii* and *Freylinia tropica*. Many of our butterflies are fond of settling and drinking at muddy puddles for the minerals they supply – an easy need to cater for. These minerals are required to produce the pheromones used in mating. Take time to admire the caterpillars too. They come ornately wrapped in a variety of brightly decorated packages that never ceases to amaze! Buy a well-illustrated reference book to learn more about these lovely and elusive creatures.

Diadem male.

It's easy to propagate your own plants

Reproducing plants, whether from seed or cuttings, is a challenge to anyone with a little time to spare. It frees you to experiment with 'slips' or seed from friends and affords you the pleasure of giving away plants grown by yourself. Plants are expensive, and one can save a lot by multiplying one's own groundcovers, bedding plants and pot plants – it's fun and easy to do.

Propagating plants from seeds

Many of our indigenous plants are easily grown from seed. This is a good method to raise large numbers of a particular plant and many are best propagated in this way. With seed and cross-pollination, each new plant is an individual and you can never be sure of the result. Patience and perseverance are required – there are no short cuts! Practise with easy things first.

Harvest fresh ripe seed from indigenous plants in your area, and sow it as fresh as possible. In general, the longer seeds are stored the fewer will germinate. Always remove seed from fruits, especially soft fleshy ones, which can rapidly become mouldy and lose their viability (fungus kills them). Throw away insect-damaged seed. Store good seed in a dry airy place, in paper bags. Sow in spring or summer when the weather is warm. In winter, when it is cold, germination is dicey, they grow more slowly and tend to rot or damp-off more easily. Plants originating in the winter rainfall area *must* be sown in autumn – March to May. Warmth and moisture must be carefully monitored for the germination period – a reasonably constant temperature is usually preferred and the soil must never dry out.

Seeds need to absorb water to germinate. In the case of seeds with hard coats (i.e. seeds with coats that are impervious to your bite), you can help by doing one of the following before sowing:

- pour just-boiled water over the seeds and allow them to stand for 24 hours until softened and swollen;
- crack or cut a small hole in the seed coat;
- rub the seeds on sandpaper until a part of the seed coat rubs off.

Certain seeds need to be scorched by fire, others to pass through the gut of some animal while many fynbos seeds need smoke. These seeds may be more difficult to grow and will require some ingenuity!

TOP TO BOTTOM:
1 Level seed mix and water well. Set out seeds.
2 Sprinkle with a light layer of clean sand.
3 Germinating seeds.
4 Seedlings almost ready for transplanting.

How to sow your seeds

Punch holes in the bottom of a suitable shallow container – it must drain well. Soggy soil will kill seedlings. Use seed-trays, margarine containers, plastic ice-cream containers or the base of a two-litre plastic cold-drink bottle. Cut off the top section, invert the bottle and insert into the base to form a protective dome. Use a fast-draining growing medium that is not prone to packing or holding water. Sterilised, commercial potting mixes specifically prepared for indoor plants are perfect. Don't use pure garden soil as it compacts and drains poorly, slowing seedling growth and causing damping-off. Never use clay soil as it holds too much water and seedlings may rot. A mix of bark, compost, coarse river sand and good, light topsoil should work well.

Fill the container with the growing medium, level the surface, and water well. Spread seeds evenly over the top, spacing them far apart. Sprinkle very fine seed like salt (not too thickly). Larger seeds can be placed by hand. Leave a reasonable space between seeds, because later you will need to dig up each young plant to replant it into a larger container. Cover larger seeds with a light layer of seed compost or clean sand (use a sieve if preferred). The finest seeds should not be covered at all. Do not bury the seeds too deeply or they won't be able to germinate! A good guide is to cover the seed with a layer of compost as thick as the seed itself.

Makeshift greenhouses

Place the seed-tray in a shady, protected place away from heavy rain and excessive sunshine, perhaps under the eaves of the house, on the east or south side. Water the seeds carefully. Use a very fine spray to avoid washing them out of the container or otherwise disturbing them. For seeds to germinate, a constantly warm and damp atmosphere is needed. People have invented all sorts of makeshift greenhouses to furnish this need, for example:

- Enclose the seed container in a clear plastic bag or even clingwrap. Do not allow the plastic to touch the surface of the soil and use elastic to keep it in place.

- Cover the tray with a sheet of glass to retain humidity. Plastic and glass prevent water from evaporating and no further watering should be necessary until the seeds have germinated. Containers enclosed in plastic must be kept out of direct sunlight – heat will build up under the plastic and kill the seeds.

- To provide moisture and warmth for a delicate cutting make a 'lid' out of an old clear plastic bag. Place a framework of sticks in the pot to support the plastic bag – do not let the plastic touch the cutting. Secure the bag with an elastic band.

- A glass jar or bottom section of a two-litre plastic cold-drink bottle can be inverted over a plastic pot of cuttings, especially in hot, dry or windy weather.

'Greenhouses' and containers do not need to cost the earth. The dome ensures constant warm, damp conditions.

A warm, damp atmosphere can be achieved in many other ways too, such as an old fish tank, covered with plastic – experiment, I'm sure you'll come up with lots more.

Check the covered seeds carefully. As soon as they germinate, remove the covering. Do not cover them again, as they will be weakened: they need light and a little sunshine to develop strong and well. Water with care to avoid washing them out of the soil. Do not let them dry out – young seedlings die easily. However, overwatering can cause damping-off. Keep them in a protected place (possibly dappled shade) for a while, until they are a little stronger. Slowly transfer them to a more and more sunny spot (this is called 'hardening off').

Transplant the seedlings into separate growing containers (7,5 cm pots or larger trays) as soon as they are robust and start to crowd one another out. This is usually about when the first pair of true leaves emerges. Take care – seedlings are frail and easily damaged. Use a broad-bladed knife to lift (or 'cut out') the seedling (or clump of seedlings), with soil attached, from the tray. Keep as much of the original soil around the rootball as possible so as not to disturb the roots too much. Some people tip the seedlings out of the tray onto the ground before transplanting, but I have found this to be a little dicey. One shouldn't tip the baby out with the bath water! The growing medium should be an exceptionally well-drained mix, for example coarse bark, about 30% light compost, some humus or leaf mould and some good light topsoil.

Avoid handling the seedlings if possible: hold by the rootball soil, or hold gently by the seed leaf. Handling the stem might cause injury and subsequent death. Firm the growing medium with your fingertips and water gently but well, to ensure that the roots are in contact with the compost. Keep them in the shade for 2–3 days to allow them to recover. Once re-established, replace in bright light and fertilise weekly with liquid fertiliser.

Propagating plants from cuttings

Not all plants are easily raised from seed, and in these cases it may be preferable to propagate them from cuttings ('slips') or even truncheons. Young plants propagated from cuttings and truncheons mature, flower and fruit sooner than seedlings do. They are also exact replicas of the parent plant so you have some quality control! Many plants are easily grown from cuttings and this method is sometimes quicker and more dependable than from seed. Leaves, stems or roots are most often used depending on the plant. Use this technique for plants that are difficult to propagate from seed, e.g. foliage plants, or plants that do not produce flowers or fruits. It is an excellent means of reproducing large quantities of a particularly successful groundcover or a favourite flowering plant such as a pelargonium or plectranthus in your garden. When the parent plant has passed its prime you will have replacements – some plants are only at their best for two to three seasons. Friends and relatives are usually thrilled to get the leftovers!

TO PREPARE AND 'PLANT' CUTTINGS

Make holes in the bottom of a suitable container – anything clean will do. Fill it with clean, coarse river sand or seed compost or fine bark, from a nursery. A mixture of the two is fine. The most important criterion for this medium is that it holds some moisture, yet drains freely.

Take cuttings in spring or summer. Those with firm tips are easier to root than hardwood sections. Cuttings harvested too early in the season may still have soft tissue and may rot. If vast quantities are to be harvested, put the newly cut material into a large, wet plastic bag. Do not allow cuttings to dry out as this reduces the chances of success. Succulents are the exception – leaves or stems should be left to dry out (callus) for a day or two before planting. Use pure, coarse river sand and keep it rather more on the dry side. Do not enclose succulents in plastic or place under a dome as this could cause rotting. Always stick to the following rules to improve the chances of success:

- The cutting should be 10–15 cm long (if the parent plant is big enough), with 3–4 nodes.

- Always make a clean cut, just below a node, using a blade, a very sharp knife or secateurs. Cuttings with jagged ends are more susceptible to disease and rot and are less likely to root.

Make a clean cut just below a leaf node.

Trim off lower leaves.

Leave two pairs of leaves at tip of cutting.

- Use a sharp knife or secateurs to strip leaves and buds off the lower half of the cutting before inserting it into the growing medium. Leave about two pairs of leaves at the tip of the cutting.
- Use a thickish stick or old pencil to make suitable holes in the growing medium – pushing cuttings directly into the soil leaves jagged ends – avoid this at all costs as they may rot.
- If desired, the cutting can be dipped in a rooting hormone. This can improve success in cooler areas.

Use a pencil or stick to make holes for the cutting.

Place the cutting in the hole in the growing medium and press the soil down firmly around the stem. Keep the soil moist, but not soggy. Use one of the 'greenhouse' options suggested above. Keep the cutting in a sheltered and shady spot and after about 3–4 weeks (some plants take longer) it should have a nice new set of roots. A mistbed improves chances of success.

Now transplant it into a slightly bigger container with good loam soil, some fine to roughish bark and lots of compost. Take care not to damage or

Insert cutting and press soil firmly around it.

disturb newly formed, delicate roots when transplanting. Feed with liquid fertiliser. A cutting or pot plant is ready for transplanting when roots appear at the bottom of its present container.

'Potting on' seedlings and cuttings

Transplant seedlings or cuttings gradually into bigger and bigger containers, until they are sturdy and strong and about 1 m tall before planting them into the garden. They may otherwise not be robust enough to handle the prevailing conditions.

It is always best to slowly increase the size of the container each time you replant a plant. If a small plant that has been growing in a tiny container is suddenly replanted into the biggest container available, the plant often deteriorates or stops growing altogether. This is because the soil becomes sour and unhealthy where there is no root activity. So, it is better to provide just enough fresh mix for the roots to grow into each time. It is therefore advisable to increase the size of the container gradually. Have fun – experiment with a few spare rooted cuttings to test if this is true!

Propagation glossary

bottom heat	Best described as a gently heated 'sandbed' in a propagation house. Some plants need a higher temperature to root. Heating is provided by a thermostatically controlled heating cable buried in the sand/propagation medium. Temperatures are usually between 23 °C and 27 °C for rooting cuttings. Cuttings are rooted directly in the medium which is 10–15 cm deep, or trays of cuttings can be placed on this bed of sand.
cold frame	Cold frames are rough shelters, often covered by plastic sheeting. They have no greenhouse facilities like heating, but are warmed by the sun's rays and can be ventilated and/or shaded. They are used to harden off transplants raised in the greenhouse or hotbed, before putting them outdoors.
callus	Protective tissue formed by a plant to cover wounds. Many cuttings need to callus before rooting.
coarse river sand	Sharp sand, lime-free and generally used in composts to improve porosity, especially for growing succulents.
cutting (hardwood)	A cutting about 15 cm long with ± 4 nodes made from the mature (previous year's) growth of a woody plant, usually in late autumn or early spring. These can vary from pencil-thick to 25 mm diameter. Make bottom cut horizontal, and just below a node. Top is cut at an angle to allow water runoff. Root in a cutting bed. Only 5 cm should be exposed.
cutting (heel)	A cutting that is pulled from a stem and left with a 'heel'.
cutting (herbaceous)	A cutting 2–5 cm long of a non-woody plant. Can be taken from the tip of the stem and rooted at any time in the season.
cutting (semi-hardwood)	A cutting 11–15 cm long made in early to late summer from the partially matured new growth (current season's) of a woody plant. Prepare cutting as for hardwood.
cutting (softwood)	These are taken in late spring from succulent new growth (tips, 6–10 cm long) produced that season. They usually root faster than hardwood and semi-hardwood cuttings, but they have to be taken at exactly the right time. If they are too tender they will rot. Make cut directly under node.
cutting (tip)	Cuttings that are taken from the tip of a growing stem.
damp off	A disease in which a soil- or airborne fungus attacks seedlings (especially young ones) at the soil line, constricting the stem and killing the plant. The seedlings quickly wilt, rot and die. To help prevent this, sterilise the growing medium before planting. Covering the seeds with vermiculite and perlite may also help.
deadhead	Removing all the dead flowerheads after a plant has finished flowering.
deciduous	Loses all of its leaves in winter.
fertiliser (inorganic)	A chemical fertiliser (e.g. slow-release 3:2:1).
fertiliser (organic)	Natural fertilisers like Kelpak, Seagro, Nitrosol, hoof and horn meal and bonemeal.
fungicide	A chemical used to control a fungus-caused disease, e.g. Kaptan, Bordeaux, Funginex, Dithane M45 and Virikop.
harden off	Getting plants gradually used to outside conditions when they have been grown under glass or in heat. Plants taken from protected conditions (houses, greenhouses, shadehouses, misting units, etc.) must be hardened off gradually. They are placed in cold frames for about 7–10 days before transplanting. Temperature and water is reduced, and there is a controlled and increasing exposure to the elements.

hardy	Plants which can tolerate a considerable degree of exposure to frost or other cold conditions.
humus	Sweet-smelling, rotted and decayed vegetable matter.
leaf mould	A rich mixture of partially decomposed leaves – an important soil conditioner.
mist spray	An automatic misting device in a propagation house, which is set to provide a fine mist at regular intervals. Without misting some cuttings experience severe water loss and may die.
mulch	A thick layer (5–10 cm) of organic material, e.g. compost, dried leaves, straw, root-free grass clippings, etc. that is placed around plants and on the surface of flowerbeds to prevent water evaporating and wind and water erosion. It also keeps the soil porous and the surface cool in summer and warmer in winter. Remember to retain a mulch-free circle around the base of stems and trunks to prevent fungal diseases.
node	The point on the stem (usually slightly enlarged) where leaves and buds arise (also branches). The *internode* is the section of stem between the two nodes.
organic matter	Any organic material (anything that has lived and can break down), e.g. dead leaves, ground bark, decomposing pine needles, sawdust, compost, manure, peanut shells, etc. that can be dug into the soil to improve its fertility and texture.
overwatering	Applying an excess volume of water at too-frequent intervals for healthy plant growth.
pinching out/back	Pinching out the growing tips (terminal bud) between thumb and forefinger, to multiply side growths (promote branching) and promote a compact habit.
pot-bound	A plant that has been in its pot too long. The over-abundance of very matted and tangled roots in the pot jeopardises the health of the plant.
potting on	Potting on or 'bagging' is the process of transferring a plant from a small container to a slightly larger one to improve its growing conditions, thereby speeding up its growth.
prick out	Removing seedlings from a tray or bed to another site where they are set out further apart to allow uninhibited growth. It is best to use a broad-bladed knife to lift or 'cut' them out of the soil in the seed tray with as much of the original soil attached to their roots as possible. Do not damage the roots or young plant.
rhizomes	A horizontal thickened swollen underground stem – a storage organ.
runners	An aboveground stem (or long shoot) produced by plants in order to multiply vegetatively. One or more plants may be attached and produce roots as they touch ground.
rooting hormone	A plant hormone (e.g. Seradix) used to promote rooting of cuttings.
semi-deciduous	Loses some of its leaves in winter.
sowing depth	= 1,5 x seed size. This means that the depth of soil used to cover the seed should never be more than one and a half times the width of the seed.
tender	A plant which cannot tolerate frost, but there are varying degrees of tenderness.
tip	The top end.
transplanting	To prevent seedlings from becoming leggy due to overcrowding in trays, they must be 'pricked out' and transferred to small individual containers.
truncheon	A 'giant' cutting! Actually a section of a branch, clean cut and not less than 5 cm diameter and 1 m long. Cut the branch off where it joins the trunk. Remove most leaves and smaller twigs, leaving a few at the top. Plant this truncheon directly into the soil where you want the plant to grow. If the plant has a milky latex or thick, fleshy bark, allow it to dry for a couple of days before planting. Place a layer of clean river sand at the bottom of the hole to encourage rooting and prevent fungus growth and rotting.

How to use this book

Choose the best shrub for a particular position

1. Study the gaps in your garden carefully and ask yourself the following questions:
   - What shape of plant would I like in that gap?
   - How big should the plant be when it is fully grown?
   - Do I want an evergreen?
   - Must the plant be frost hardy?
   - Will the plant be in the sun or shade?
   - What colour flowers do I want?
   - When should they bloom?

2. Once you have decided on the kind of plant you would like, e.g. a small shrub, turn to that section of the book and use the symbols to help you make your selection.

A guide to the frost areas of South Africa

Every garden has a number of microclimates, allowing the gardener to grow a wide variety of plants. For example, some of the more frost-tender ones may thrive against a north-facing wall warmed by the winter sun and offering protection from cold winds. Similarly, trees provide shelter for more delicate, shade-loving plants planted beneath them.

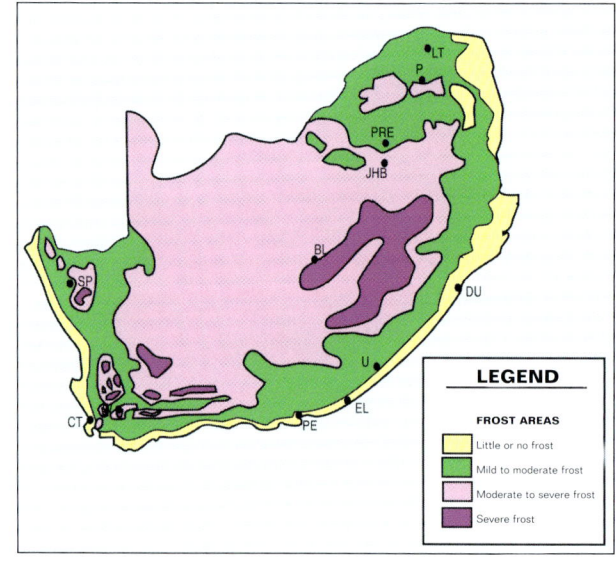

Note that a mature plant indicated as half-hardy may be burnt by frost during an average Pretoria winter, but will recover in spring. A younger plant of the same species could, however, be killed by the same frost, and would therefore require some protection.

The map gives a rough indication of the frost areas of South Africa. But like a garden, a town also has different microclimates, so when starting a new garden or moving to a new area, consult your neighbours or local nursery.

SIZE

The size of a plant is indicated as follows: **1,5 m × 1,8 m**

Where 1,5 m is the height, which is always given first, and 1,8 m is the width. Always allow enough room for a plant to spread – its natural shape will be spoilt if it is crowded.

Key to symbols

CROWN SHAPE (TREES)

- Rounded, dense
- Bare stem, small rounded crown of leaves and/or branches
- Rectangular, dense
- Weeping
- Rounded – sparse, open texture
- Long narrow – sparse, open texture
- Contorted
- Flat to umbrella-shaped
- Conical
- Low-branching, rounded, dense

SHAPE (ALL OTHER PLANTS)

- Rounded
- Clump of strap-shaped leaves
- Clump of stiff, leathery or succulent leaves
- Tree fern
- Bulb or corm
- Oval crown, base bare and untidy
- Ground cover
- Palm or banana-like
- Climber
- Aquatic or water-loving

GROWTH REQUIREMENTS

- Requires little water
- Requires moderate water
- Requires lots of water
- Requires full sun
- Requires semi-shade
- Requires full shade
- Frost hardy
- Half hardy
- Frost tender

CHARACTERISTICS

- Deciduous (loses leaves)
- Evergreen
- Attractive flowers
- Succulent (cactus-like)
- Wind resistant
- Drought resistant
- Waterlog resistant
- Attracts birds
- Attracts useful insects, e.g. bees and butterflies

How the information about each plant is presented

102 *Large shrubs*

Pistol Bush ■ Pistoolbos, isiPheka (X), iHlwehlwe (X), uHlwalana (Z), Lothabe (Z)

Duvernoia adhatodoides ACANTHACEAE

DESCRIPTION AND USES
The Pistol Bush forms an attractive, usually rounded shrub with large dark-green leaves and showy fragrant white flowers streaked red. Under favourable conditions (warmth and moisture) it may grow into a small tree up to 6 m high. Brown velvety club-shaped fruits burst open with a loud explosive sound, hence the name 'Pistol Bush'. Carpenter bees apparently pollinate the flowers.

PROPAGATION
Propagate it from seed. Under good conditions in the garden it seeds itself freely – give a plant to a friend.

CULTIVATION
A wonderful shrub for those shady moist parts of the garden, and beautiful in flower. It is easily grown, has pretty foliage, and is perfect for home gardens (small to large), school gardens, hotel grounds, parks and office gardens, in warmish areas. It grows well in Pretoria, but in exposed positions can sometimes be burnt by frost. When the frost is over, cut back the dead leaves. Position it under large trees, or use on the shady south side of the house, or to decorate a shady entrance area. Try planting it next to a stream or dam. Always ensure that it has some shade and sufficient moisture, or it will not do well and the leaves will turn yellow. This fast-growing plant prefers light well-drained fertile soil – add plenty of compost, mulch well, and water regularly for best results. Feed with slow-release 3:1:5 fertiliser at intervals of 6–8 weeks throughout the growing season. Prune if necessary to shape and neaten. Do this after the plant has flowered, unless you want to harvest the seed for propagation purposes.

NATURAL DISTRIBUTION
Coastal forest, forest margins and along streams. NTN 681

RELATED SPECIES
Duvernoia aconitiflora (Lemon Pistol Bush, Geelpistoolbos) is a fast-growing and easy-to-grow bushy rounded shrub (3 m x 3 m) with masses of two-lipped, pale lemon-yellow flowers. Sunbirds visit the flowers regularly. Cultivation requirements are similar to those of *D. adhatodoides*, except that it prefers lighter shade or slightly more sun. It is a little hardier and does not need as much moisture. NTN 681.2

D. aconitiflora

Summer

3,5 m × 3 m

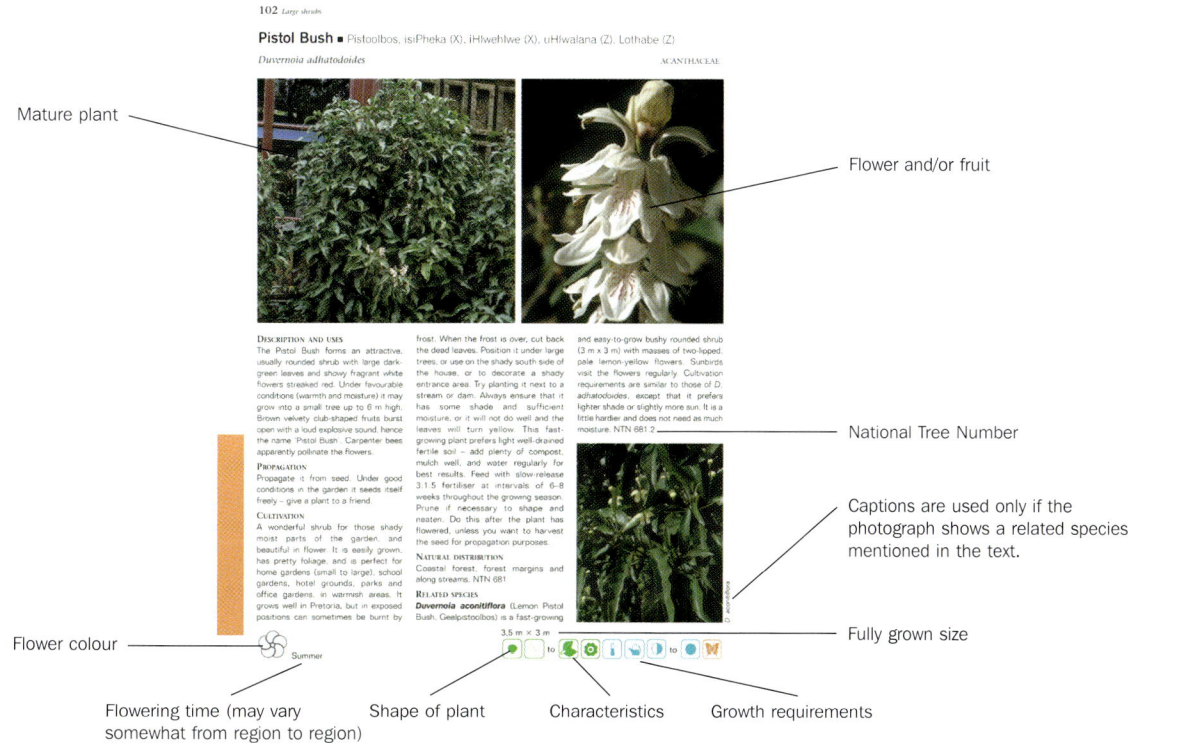

Impala Lily ■ Sabie Star, Impala-lelie

Adenium multiflorum　　　　　　　　　　　　　　　　　　　　　APOCYNACEAE

Description and uses
When Impala Lilies flower in the Kruger Park, I would love to take one home! The beautiful white flowers are star-shaped, each petal tinged pink and heavily outlined with red. The broad oval leaves are dark glossy green above and paler below. These plants have enormous tuberous underground stems that allow them to withstand drought and heat. They like a very hot summer and a dry, temperate winter. The aboveground trunks and branches are smooth and swollen (succulent), developing into the oddest shapes and sizes – a little like miniature baobabs. They grow exceptionally slowly and are leafless for a good part of the year.

Propagation
Propagate from seed or cuttings. Take cuttings after flowering time, but before leaves appear. Root in clean, coarse river sand or sandy soil. Water very sparingly. Sow seed (September) in a well-drained mix of sandy loam, coarse river sand and coarse compost (equal parts). Place in container and level. It is essential that water drains very freely from the container – seedlings damp off (rot and die) easily. Sow seed and press down lightly. Cover with fine river sand (depth 1,5 x seed size). Germination time, 4 days. Expose seedlings to sun as soon as possible to strengthen them. Prick out carefully into individual small (about 500 ml) pots/bags when they are 2–3 cm in height. Young plants can grow to about 8 cm high in the first season. They stop growing in winter and should only be watered very little. One-year-old plants can be planted into the ground where the climate permits (hot, frost-free areas). The end of July is a good time, after the dormant period and before the leaves appear. Alternatively, transplant into slightly larger pots.

Cultivation
Don't allow your Impala Lily to grow too large in the pot – the underground stem may become too big or stunted. Never water this plant in winter. Always err on the side of too little water, rather than too much. Allow the soil in the pot to become very dry before you water again. This plant happily establishes itself in gardens with a suitable climate (Louis Trichardt, Musina). Don't try to grow it in cold climates – it is used to excessive heat and prolonged droughts. If grown in gardens that are warm with only occasional frosts, plant in a very warm sheltered spot. Perfect for a rockery, grouped with other interesting succulents like aloes.

Natural distribution
Hot dry areas, in sandy soil, in open sunny positions. NTN 647.3

> *HINT: Before sowing, it might be advisable to treat the soil with a fungicide to help prevent damping off. Use Kaptan (5 level teaspoons to 5 litres of water). Drench seedpans and soil thoroughly before sowing. Allow soil to dry a little.*

 Jun-Aug

1,8 m × 1,5 m

False Buchu ■ Basterboegoe, Bosboegoe, Valsboegoe, Wildeboegoe, Buchu (Khoikhoi)

Agathosma ovata 'Kluitjieskraal'   RUTACEAE

Description and uses
A neat and charming evergreen that is upright to spreading, with small dark-green aromatic leaves that closely 'hug' the stem. Tiny pink star-shaped flowers are freely produced in winter and attract masses of bees. *Agathosma* has a distinctive scent that intensifies as one brushes against the plant or bruises the leaves. The oil has medicinal properties and also acts as an insect deterrent. False Buchu is used only to a limited extent for medicine and essential oil. The flowers are popular for flower arrangements.

Propagation
Propagate from seed or cuttings. Harvest seed at precisely the right time. If collected before it is properly ripe, it won't germinate; wait too long and the capsules burst open explosively, shooting your seeds in all directions. It seems best to collect the seed when the first capsules start to open. Place unopened capsules in a paper packet or box to dry and ripen properly. In a plastic bag, they remain moist and may rot. Sow fresh seed (Feb–Mar) in a sandy loam and bark mix (2:1), 2 mm deep. Place in a cold frame in light shade. Keep moderately moist. Germination takes 4 weeks. Success rate is 70%. Collect softwood or herbaceous cuttings from an actively growing mother plant (Sept–Nov). Treat with Seradix 3 and a fungicide. Root in a bark and polystyrene mix (3:1). Place in a mist unit, bottom heat 23 °C. Rooting takes 8–11 weeks; hardening off, 2 weeks. Success rate, 70%.

Cultivation
Plant out between autumn and early spring, after the first rains. Don't disturb the roots of buchus when transferring from bag to ground. Do not add inorganic fertilisers to the planting hole. The plant prefers acid, well-drained humus-rich sandy loam. Some people prefer to plant it without compost, but it performs better if compost is added. Place in full sun at the coast. Inland, it tolerates light shade and likes plenty of water in winter – don't let it dry out completely in summer. Mulch well and water moderately for the first summer, until well established. Growth rate is fairly slow and the roots are non-invasive. Perfect as a filler, a clipped hedge or as a groundcover. Use it in a pot on a patio, or train it to trail over the edges of a hanging basket or wall. It is important to prune and shape these plants correctly when young. Pinch back new growth, and prune lightly at the end of the flowering season to encourage a bushy, compact growth with more flowering stems. Do not allow a faster growing shrub to overgrow a buchu, or it will lose its shape forever. Prefers temperatures from 5 °C to 25 °C and has a life expectancy of 10–15 years.

Natural distribution
Mainly on mountain slopes (Ceres area).

Related species
Agathosma betulina (Round-leaf Buchu) and **Agathosma crenulata** (Oval-leaf Buchu) are grown commercially for their medicinal properties and valuable essential oils.

80 cm × 1m

May–Aug

Basuto Kraal Aloe ■ Basoetokraal-aalwyn, inHlaba empofu (Z), iKhalene (Z), inTelezi (Z)

Aloe tenuior ASPHODELACEAE

Description and uses
This fast-growing, free-flowering aloe forms a dense, rounded shrub with succulent grey leaves. Spikes of delicate red flowers with yellow tips cover the bush in spring and autumn. A beautiful yellow-flowered form is also available. The flower colour is naturally very variable, and the rich nectar attracts bees and both Whitebellied and Black Sunbirds. The Xhosa used this plant to build kraals to protect their cattle. Traditionally, leaves are used as protective charms, and a root and leaf remedy treats tapeworm infestations.

Propagation
Easily propagated from cuttings, which must be kept fairly dry to prevent rot. Remove cuttings (or offsets) with a sharp knife/secateurs. For *A. tenuior*, cut cleanly just below a leaf node. Root in the same mixture used for sowing seed (see Hint). Insert stems or rest offsets on mixture and fill in the top 2 cm around cuttings/offsets with pure river sand. Rooting hormone will speed up the process.

Cultivation
Plant the Basuto Kraal Aloe in well-drained, compost-enriched soil and water well in summer; keep dryish in winter. Prune back by about a third to a half in autumn to keep it neat and to encourage bushiness. Don't throw away the prunings, rather use them to quickly increase the width of the bush. Insert them directly into the soil at the base of the bush so that it can eventually develop into a nicely rounded, dense shrub. Position this aloe strategically between large rocks on a natural rockery, or plant it in a shrub border or succulent garden. Use it in a small townhouse garden where its bright flowers can provide colour for many months of the year. Not really prone to pests – sometimes aphids on the flowers.

Natural distribution
Rocky areas in South Africa and Lesotho.

HINT: Aloes also grow well from seed, but are inclined to damp off if overwatered. To collect aloe seed, place an old stocking/pantyhose over the whole seed head when the first capsules/fruits begin to split open. When most have split, cut the seed head off and leave it in a warm place to dry thoroughly. Sprinkle seed evenly over a mixture of equal quantities of coarse river sand, compost and sandy soil. Ensure that water drains freely from the container and mixture. Cover seed with a layer (depth 1,5 x seed size) of small lentil-sized gravel or pebbles. This helps to prevent damping off and keeps the seedlings upright. Place in a warm, shady spot and keep moist. A weak Bordeaux solution added to the water may also help to prevent damping off. Handle seedlings with great care, as they have delicate roots. Prick out and transfer to larger trays when they are about a year old. Some of the more robust species may be transplanted into the garden after a year.

 Aug–May, sporadically throughout year

1,5 m × 1,5 m

Pink Mallow ■ Pienk-kiesieblaar, Harigemalva
Anisodontea scabrosa　　　　　　　　　　　　　　　　　　　　　　　　MALVACEAE

A. julii

Description and uses
Pink Mallow is a rewarding free-flowering shrub with soft hairy three-lobed leaves that are aromatic and often sticky. Erect and rather sparse in habit, it is softened by masses of smallish (about 2 cm diameter) pink-mauve flowers, each streaked with dark-pink markings.

Propagation
Propagate it from seed or from cuttings.

Cultivation
It grows fairly fast in any soil, but add plenty of compost for better results. Apply a thick mulch layer and replenish regularly. Water well in summer, less in winter. Fertilise with slow-release 3:1:5 at intervals of 6–8 weeks throughout the growing season. Prune the plant back by about a third in autumn to keep it neat. Always pinch out the growing tips of *Anisodontea* species to encourage bushiness otherwise they tend to become a little leggy, sparse and untidy. Mass plant to form a good groundcover, place it in an informal shrub border, or use as a temporary filler until slower shrubs have grown and established themselves. Perfect for 'prettying-up' a sunny entrance area or bed on the north or west wall of the house. Experiment with it as a pot plant for temporary colour on a paved area. Useful for tiny townhouse gardens, where it could easily be replaced should you wish for a change. In a small garden, use dainty shrubs and leave open space to prevent clutter, allowing your plants to be seen to advantage. Pink Mallow is now available at nurseries as a standard. Suitable for a fynbos and succulent karoo garden, but also performs well on the Highveld – experiment with it in other areas of the country.

Natural distribution
Along the coast; also in thicket and grassland.

Related species
Anisodontea julii subsp. julii is a rather sparse shrub (about 2 m × 1 m) with soft velvety foliage and attractive pink-mauve flowers (Feb–Mar) streaked with red. It may occasionally form a small tree up to 4 m high. Cultivate as for *A. scabrosa*.

> **HINT:** *Plants grown from seed may not resemble the parent exactly. Propagate from cuttings if you want to maintain a particularly good form of the plant in your garden.*

1,8 m × 1 m

 to

Aug–Mar

Grey Bush Violet ■ Grey Barleria, Grysbarleria, Fluweelblaarbosviooltjie

Barleria albostellata ACANTHACEAE

DESCRIPTION AND USES
The Grey Bush Violet has attractive woolly grey foliage and bears clusters of tubular white flowers.

PROPAGATION
Seed or cuttings. Seeds burst explosively from capsules when ripe. To harvest successfully, collect capsules when they turn brown and place in a paper bag. Level a mixture of good loam, compost and coarse river sand (2:1:1), sow seed and press in fairly firmly. Cover lightly with fine river sand. Place in 40% shade and keep moist (use a very fine spray). Germination time, 7 days. Transplant when 20 cm high. Take cuttings 12 cm long from the previous year's growth. Treat with Seradix 2 and root in clean river sand. The ideal temperature for root formation is 10–25 °C; use bottom heat if available. Plant rooted cuttings in a good soil mix and place in 40% shade. Harden off gradually. Water young plants well after planting out, until they are properly established.

CULTIVATION
Plant this dense, rounded shrub in groups of 3–5 in light shade under trees, or along an informal border, in good well-drained soil, adding plenty of compost and other organic material. It could be used to line a natural pathway in a 'woody, wildish' area – allow enough room for spread to avoid blocking the pathway. Water well in summer, less in winter. Fast-growing and naturally neat, this plant does not require too much pruning. Should it become untidy, it can be cut back by a third or more to improve the shape and stimulate new growth. Under tropical and subtropical conditions it is evergreen, but in cooler areas it may be semi-deciduous to deciduous. Deadhead the bush to keep it neat.

NATURAL DISTRIBUTION
Woodland areas of South Africa and Zimbabwe.

HINT: *Always provide plants with a nice thick layer of mulch. Mulch protects both the soil and the roots, keeping them cool and moist – encouraging better growth while simultaneously decomposing to form nutrients which are returned to the soil for the plant's benefit.*

 Sept–Mar

1,5 m × 1,5 m

  to   to

Bush Violet ■ Bosviooltjie, iDololenkonyane (Z)

Barleria obtusa

ACANTHACEAE

Description and uses
The Bush Violet is a rounded bushy shrub that tends to sprawl or scramble. It has small soft oval leaves covered in silky hairs, and masses of tubular blue-mauve, pink or white flowers. In Nelspruit I watched Collared Sunbirds sipping nectar from the flowers. The leaves are often browsed by stock and game; flowers attract butterflies.

Propagation
Easily propagated from seed or from cuttings.

Cultivation
The Bush Violet is a wonderful choice for an informal area of the garden where it can happily scramble – or perhaps, more accurately, lean – into surrounding shrubs. Plant in large numbers, in light shade under trees, to form an excellent groundcover. Experiment with this fast-growing plant in a big tub or container, where the foliage and flowers can hang over the edges. Use it to line pathways or driveways – cut back any branches that sprawl onto the path or roadway. In cooler areas, plant it in a protected position in good well-drained soil, adding plenty of compost, and water well in summer. Apply a thick mulch layer and replenish it regularly. Feed with slow-release 3:1:5 fertiliser at intervals of 6–8 weeks throughout summer. Pinch out young shoots to encourage bushiness; prune back hard after flowering. Suitable for a bushveld-type garden, but this plant seems to grow well in many other parts of the country too.

Natural distribution
Rocky koppies, thicket, bushveld and in grassland.

Related species
Barleria pretoriensis is a particularly beautiful plant, covered in masses of pale blue-mauve flowers in autumn. It is a pity that no serious attempt has been made to bring it into cultivation, though young plants are occasionally available.

B. pretoriensis

1 m × 1 m

Mar–Apr

Small Bush Violet ■ Bosviooltjie, Kleinbosviooltjie

Barleria repens　　　　　　　　　　　　　　　　　　　　　　　　　　　　ACANTHACEAE

Description and uses
The Small Bush Violet has somewhat shiny, soft green leaves and deep purple-mauve or pink-red tubular flowers. It usually forms a rounded to spreading bushy shrub, but will scramble into surrounding vegetation if given half a chance. Flowers attract carpenter bees (*Xylocopa* species) and other insects.

Propagation
Easily propagated from cuttings. An even easier method is to lift rooted runners – plant into individual containers (size depends on size of roots attached) and water carefully until properly established. Don't let them dry out completely, but don't keep them saturated. When growing strongly, transfer to the garden.

Cultivation
Fast-growing and wonderfully easy-going, the Small Bush Violet will adapt to a number of situations. Mass plant it in partial shade under trees to form a groundcover, or plant along the edge of an informal border, or in a lightly shaded rockery. When planted in very deep shade it tends to become lanky and untidy and does not produce as many flowers. It quite happily scrambles up into nearby trees and shrubs (up to 2 m). New branches tend to take root as they touch the ground, so this plant can quickly increase its territory if not kept under surveillance as my garden will testify! Plant it in a large container, or on top of a low wall, where its foliage and flowers can cascade down and show off to advantage. Provide good, light, well-drained soil, and plenty of compost and other organic material. Spread a layer of mulch on the surface of the soil after planting, and renew regularly. Water well in summer but quite a bit less in winter. Plants thrive when fed with slow-release 3:1:5 at intervals of 6–8 weeks in the growing season. Pinch out young shoots to encourage bushiness; prune back hard after flowering at the end of summer/winter to keep it neat. The prunings should be regarded as free mulch!

Natural distribution
Woodland and forest, from KwaZulu-Natal northwards to tropical Africa.

Related species
A compact (50 cm × 50 cm) cultivar called **Barleria repens 'Rosea'**, with beautiful rose-pink flowers, is available.

B. repens 'Rosea'

Feb–Apr

75 cm × 1 m

Thorny Bush Violet ■ Spiny Yellow Barleria, Yellow Barleria, Geelbarleria

Barleria rotundifolia　　　　　　　　　　　　　　　　　　　　　　　　　　　　　ACANTHACEAE

DESCRIPTION AND USES
A spiny sprawling shrub with small glossy-green leaves, decorated by attractive yellow tubular flowers that look like tiny faces with tongues hanging out!. The flowers attract insect pollinators and butterflies (e.g. Yellow Pansy), in turn attracting insectivorous birds.

PROPAGATION
Seed or cuttings. See *B. albostellata* (page 36) for tips on collecting and sowing seed.

CULTIVATION
The Thorny Bush Violet grows moderately fast and prefers fertile well-drained soil so add plenty of compost, and water it regularly in summer. Apply a thick mulch layer and replenish constantly. Feed from time to time with slow-release 3:1:5 fertiliser in growing season. Use this plant to form a natural spiny barrier to discourage unwanted traffic in problem areas. If necessary, prune to shape and neaten. Set plants out fairly close together (about 40 cm apart) to ensure a more solid boundary. Position it in groups of 3–5 in light shade under trees, or display clumps on a large natural rockery on a farm or a nature reserve. A group of plants looks good in a shrub border or along a lightly shaded pathway. Experiment with a few plants in large tubs or pots on a patio or pool deck. The potting mix must be very well-drained – use half to three-quarters by volume of coarse light compost and the rest can be made up of good loam, and even a small amount of coarse river sand. Add about 250 ml bonemeal or superphosphate per wheelbarrowful of mix. Particularly suited to bushveld gardens, it also does well in Pretoria, Kloof and other areas.

NATURAL DISTRIBUTION
Rocky koppies and hillsides in the Lowveld.

HINT: *Ensure that plants in containers are fed and watered regularly, otherwise you will not have much success with them.*

1 m × 1 m

   to

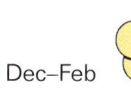

Dec–Feb

Pink Joy ■ Plakkie, Beestebul, Karkey (Khoikhoi), Thlakeni (So), umXhalagube (X)

Crassula ovata (= *C. portulacea*)　　　　　　　　　　　　　　　　　　　　　CRASSULACEAE

DESCRIPTION AND USES
The Pink Joy is a large, rounded shrub with dark-green succulent leaves, coloured by masses of pretty pale-pink, star-shaped flowers in winter. The flowers attract a variety of insects, such as bees, beetles with a metallic sheen, and beautiful butterflies (e.g. Burnished Opal, Cape Black-eye and Pale Hairtail). The roots were apparently once eaten by the Khoikhoi peoples. Other African tribes are said to have grated the root and cooked it, after which it was eaten with thick milk. The leaf juice has been used as an astringent.

PROPAGATION
Use cuttings, which must be kept fairly dry to prevent rot. Root them in clean, coarse river sand. Remember that succulents must be left to dry out for a couple of days, or even a week, before putting them into the rooting medium. This plant grows easily in any well-drained soil – add compost and bonemeal, and water sparingly in summer. Provide a mulch, and fertilise with slow-release 3:1:5 at intervals of 6–8 weeks throughout the growing season. In colder frosty areas place the shrub in a sheltered position to prevent damage to the flowers. It is wind-resistant, fast-growing and tolerant of coastal conditions; ideal for a rockery or an informal shrub border in hotter, drier areas. Grow it in a large decorative container and place on a patio or on the paving next to a garage. After flowering, cut off the dead flowerheads to neaten the bush. If desired prune and shape a little, but this will not usually be necessary. Use the prunings to multiply the plant – try planting directly where you need them, but do not overwater. Suitable for noorsveld and fynbos gardens, it also performs well in many other areas of the country, like Pretoria.

NATURAL DISTRIBUTION
Rocky slopes, often in sheltered ravines, or intermingled with riverside vegetation.

> **HINT:** *South Africa is blessed with an unbelievable variety of succulents (nearly half the world's species are found here). Use Plakkies and other succulents to create an unusual and striking low-maintenance garden in suitably dry areas.*

 May–Jul

1,8 m × 1,5 m

Dune Marigold ■ Duinegousblom, Seegousblom, Perdeblom
Didelta carnosa ASTERACEAE

Description and uses
A grey-green to silvery rounded bush that is usually abundantly covered in golden-yellow daisy flowers. Its round bushy habit and attractive foliage (silvery-grey leaves that look almost succulent) make it a welcome addition to the garden. The Duinegousblom is particularly well suited to conditions along the south-western Cape coast and in Namaqualand, but also copes fairly well in other areas. It is often closely planted like a groundcover to control sand, and happily faces strong winds and salty seaspray. This plant is spectacular in Namaqualand and the Richtersveld, where small rounded yellow bushes dot large tracts of land. I have not seen them inland, so cannot vouch firsthand for how well they would do there. Stock are partial to this plant and happily feed on it in winter as well as in summer when the leaves are dry.

Propagation
Propagate from seed, or by lifting side shoots that have already rooted. Plant these in pots until they are growing strongly.

Cultivation
At the coast (and elsewhere!), speed up growth by adding lots of compost to the planting holes and water fairly regularly. Experiment with this plant in a new garden for almost instant colour. Try planting some in a large pot, or along the edge of a pathway. They might also look good lining the front of an informal border. I suspect that they would be at their best for about 2–3 years, after which they may need replacing.

Natural distribution
On sand dunes, in Namaqualand, the Richtersveld, SW Cape.

Related species
Didelta spinosa (Perdebos, Slaaibossie, t'Arda [Nama]) is an erect, sturdy well-branched shrub that sometimes reaches 2 m. It has bright green leaves (with or without tiny spines on margins) and boasts masses of deep yellow to orange daisy flowers (Jun/Jul). Widespread in Namaqualand, it should make an interesting addition to gardens in that area, and possibly elsewhere. Seed ripens in September (not all mature). Sow the seed in April in seedbeds or trays. Cover lightly with sand. The seedlings can be transplanted to the garden/farmlands in May–July when about 10 cm high. They prefer sandy well-drained soil. Add compost to the planting area. Water carefully until they are firmly established. This plant has no taproot, only lateral roots that reach a depth of about 4 m. It is palatable to animals, especially mules and horses, hence the name Perdebos. Farmers can plant it on farms as a type of backup fodder plant. In early winter, animals actually prefer to eat other more palatable plants in the area, but when these have all been defoliated, they will eat the less palatable Perdebos. In summer, they are particularly fond of the dry leaves.

1 m × 1 m

Late spring, summer

Yellow Wild Iris ■ Peacock Flower, Poublom, Uiltjie, isiQungasehlati (Z)

Dietes bicolor

IRIDACEAE

DESCRIPTION AND USES
Masses of lovely lemon-yellow, spotted black flowers are displayed on tall (about 90 cm high) stalks above clumps of light-green strap-shaped leaves, each nearly 1 m long and softly drooping at the ends. It attracts bees and other insects, which in turn attract insectivorous birds (e.g. Southern Boubou, Crested and Blackcollared Barbets and robins). *Dietes* roots are traditionally used as charms to safeguard and strengthen the wearer.

PROPAGATION
Easily propagated from seed, or by dividing up large clumps, which spread by means of rhizomes. Due to the very hard seed coat, seed germinates erratically over an extended period of time (4 or more years). Sow ripe seed in a mix of compost, river sand and loam (equal parts). Use deep seed trays. Place in the shade and keep moist. Transplant seedlings into black bags to grow sturdy before placing in the garden. See *D. grandiflora* (opposite) for detailed instructions on splitting clumps.

CULTIVATION
This iris grows in any soil, but for best results add plenty of compost and water regularly throughout the year. Feed with slow-release 3:1:5 fertiliser at intervals of 6–8 weeks throughout summer. It can tolerate very light shade, but seems to do better in the sun. A wonderfully reliable plant, now extremely popular with landscapers for that very reason. It has been mass-planted in many large office complex gardens, also in schools, at shopping malls, in parking lots and at petrol stations. Plant the fast-growing, drought-resistant Yellow Wild Iris in large groups, or singly as an accent plant at the edge of steps or next to a pond. Use it to line a driveway or pathway, or to fill a whole bed, as a groundcover. Mass plant it, combined with *Agapanthus* or *Tulbaghia violacea*, for a stunning display. The options are limited only by your imagination – experiment and have some fun with this easy-going plant!

NATURAL DISTRIBUTION
Near streams and marshy places.

Oct–Jan

1 m × 1 m

Fairy Iris ■ Large Wild Iris, Feë-iris, Groot Wilde-iris, isiQungasehlati (Z)

Dietes grandiflora IRIDACEAE

DESCRIPTION AND USES
Large, delicate and rather beautiful white-orange-and-mauve flowers are displayed on long, slender stalks about 1 m tall above clumps of stiff strap-shaped dark-green leaves. They attract lots of bees and other pollinators.

PROPAGATION
Easily propagated from seed sown in September, or by dividing up large clumps, which spread by means of rhizomes. Approximately every four years, clumps can be divided to keep plants healthy and vigorous and to maintain successful flowering. Do this in autumn after flowering or in early spring, before vigorous growth begins. Two forks placed back to back in the centre of the clump can be used to pry the plant apart. Cut back older sections of rhizome. Trim foliage down by about half. Replant immediately and water thoroughly. Thereafter, keep moist until well established. See *Dietes bicolor* (opposite) for detailed instructions on how to propagate from seed. Remember to add lots of compost and other organic material to the soil when you set the young plants out.

CULTIVATION
If plants start to look a little scrappy, e.g. at the end of winter, cut off all the leaves at ground level, mulch well and feed with slow-release 3:2:1 fertiliser. Water well and they should soon recover. This plant is slightly more tolerant of drought and poor soil than *D. bicolor*. It is also able to tolerate slightly more shade. Mass plant in partial shade under trees to form a groundcover (looks stunning when all the plants are covered in their delicate white flowers), or use singly as an accent plant next to a pond, steps or an attractive rock. Called a 'Fairy Iris' because the fragile white petals not only look like fairy wings, but also have a tendency to disappear mysteriously overnight! Now a popular landscaping subject, owing to its reliability and hardiness, it is mass planted in large office complex gardens, in parking lots, at shopping malls, petrol stations and schools. This vigorous grower is an excellent candidate for poorly drained boggy areas in full sun. Combined with a mass planting of *Agapanthus* or *Tulbaghia*, it can look superb.

NATURAL DISTRIBUTION
Forest margins and near streams.

1 m × 1 m

 to

Nov–Jan

Heaths ■ Ericas, Heide

Erica species ERICACEAE

E. versicolor

E. bauera

E. verecunda

E. scabriuscula

DESCRIPTION AND USES
Ericas all have tiny fairly densely packed needle-like leaves. Flowers vary in shape from long and tubular to bell- or lantern-shaped. Ericas with tubular flowers attract sunbirds such as the Orangebreasted Sunbird – some even have flowers that are strongly curved to match the beak shape of this bird. Bird-pollinated ericas tend to have thicker, stronger stems to provide birds with a steady perch. Erica flowers are filled with nectar, and about 80% are thought to be insect-pollinated. They come in virtually all colours of the rainbow and are best seen to be believed: visit your local nursery to find out what is available – I'm sure that you'll be tempted to take one of these tiny plants home for your patio or garden. Plants range in size from extremely small to about 3 m high.

PROPAGATION
Difficult – rather purchase plants from your nursery. Cuttings (fairly new tips, 2–5 cm long) are the easiest method. Strip the leaves from the base, but don't damage the bark. Heel cuttings can also be used. Dip the base in rooting hormone, a fairly weak solution, or it may damage the bark. Root in a mix of sand and sifted humus (7:3), or granulated pine bark and polystyrene chips (7:3). Use a subdivided tray and plant a cutting in each 'holder'. Transfer to individual small pots when roots have developed. Water with a fine mist spray, never let the leaves dry out. Keep plants wet, good drainage is essential. Ericas grown this way are more robust than seedlings of a similar age. They can also be transplanted within 6 months and usually flower a year earlier.

CULTIVATION
Ericas like an airy, preferably sunny position, exceptionally good drainage (a raised bed, a slope or a rockery is perfect), and light well-drained acid soil (pH about 5) that contains plenty, but really plenty of compost (half soil, half compost)! Prepare the beds thoroughly, otherwise your plants will not survive. Ericas do not like clay soil, manure or artificial fertilisers, so avoid these. Do not disturb the root area by digging; apply a thick layer of mulch to the soil surface and replenish regularly. In the dry interior it is necessary to water them well winter and summer – do not let them dry out or become waterlogged. Cut off dead flowers to neaten the bushes, or trim lightly from time to time, if necessary.

NATURAL DISTRIBUTION
There are about 625 species of *Erica* in the southern Cape alone, and a few others along the escarpment and elsewhere.

RELATED SPECIES
Examples of plants that are available: **Erica bauera** (Bridal Heath; 1,5 m x 1,5 m) is widely grown, producing pink and white flowers for most of the year; it grows easily from seed or cuttings. **Erica scabriuscula** (1 m x 1m), **Erica versicolor** (tall erect bushes, 3 m x 2 m), flowers red and greenish yellow, long-lived and fairly easy to cultivate and **Erica glandulosa**.

Variable (depends on species)

Height/width variable

Wild Rosemary ■ Cape Snowbush, Kapokbossie, Kapokbos, Wilderoosmaryn

Eriocephalus africanus ASTERACEAE

Description and uses
A well-branched, 'furry-looking' grey-green shrub, with soft foliage that is usually covered in masses of tiny delicate white flowers or puffs of cottonwool-like fruits. The fruits are said to resemble snow, hence the Afrikaans name 'kapok'. They were used to fill pillows, or to line nests! Wild Rosemary is an unusual and beautiful shrub, especially when in full flower. The tiny, hairy semi-succulent needle-like leaves are aromatic – when crushed they emit a distinct 'herby' smell. They can be used as a substitute for rosemary, to flavour meat. Animals readily eat this plant. This is one of our well-known traditional medicinal plants (diuretic), and has been used to treat oedema (dropsy) and stomach-ache.

Propagation
Sow seed in March, or propagate from cuttings.

Cultivation
Plant Wild Rosemary in well-drained soil, adding plenty of compost, and mulch well. Feed it from time to time (growing season) with small amounts of slow-release 3:1:5 fertiliser. Water reasonably throughout the year. It does not seem to need too much pruning to keep it neat. If you plant it along a pathway, remember to allow enough room for it to spread comfortably, otherwise energy will be wasted pruning it. Set this fast-growing and hardy shrub out in groups of 3–5, or mass plant on a sunny bank to form a groundcover. Use it to line the front of a shrub border or along a pathway. Suitable for new gardens, where it can be planted as a temporary hedge or screen while slower plants are establishing. Perfect for the top of a low wall, or in a large container on the patio or at the entrance where passers-by can bump it and smell Namaqualand! Suitable for

a fynbos, strandveld and succulent karoo-type garden, but it also performs well on the Highveld – so be brave and experiment with it in other parts of the country too. It does not seem too prone to pests.

Natural distribution
Throughout the Cape, but especially near the coast.

1,75 m × 1,75 m

Jul–Oct

Golden Daisy ■ Gouemargrietjie, Wolharpuisbos

Euryops pectinatus

ASTERACEAE

E. speciosissimus

E. speciosissimus

DESCRIPTION AND USES
Free-flowering and fast-growing, this attractively rounded shrub has soft silvery-grey foliage and bright sunny-yellow daisy flowers. These flowers entice bees and other pollinators, so this plant will be a useful and low-maintenance addition to the 'bird' garden.

PROPAGATION
Propagate it from seed or from cuttings – cuttings will produce a quicker result.

CULTIVATION
Set the Golden Daisy out in groups of 3–5, or place it along the edge of a shrub border. Use it to line pathways and driveways, or as a temporary filler until slower shrubs have established. Position a single bush where it can be the centre of attraction, e.g. in a small garden. Always plant this shrub in a sunny position. It tolerates any soil type, but performs better if it is light and well drained and contains plenty of compost. Apply a thick mulch layer and replenish regularly. Feed from time to time (growing season) with small amounts of slow-release 3:1:5 fertiliser. Water well in winter, remove dead flowerheads after flowering, and prune the bush back lightly. Plants are past their best after a few years – replace them every 4–5 years.

NATURAL DISTRIBUTION
Rocky mountains and cliffs.

RELATED SPECIES
Euryops speciosissimus (Clanwilliam Euryops, Harpuisbos) flowers in spring, is of a similar size to *E. pectinatus*, and has similar cultivation requirements. It has finely divided, soft green leaves and large yellow daisy flowers. It tends to be rather bare at the base, so plant it behind other lower-growing plants like *Felicia amelloides* to conceal this.

Jun–Oct

1 m × 1 m

Honey Euryops ■ River Resin Bush, Heuningmargriet, Rivierharpuisbos

Euryops virgineus

ASTERACEAE

DESCRIPTION AND USES
Fast-growing and extremely hardy, the Honey Euryops has fine dark-green needle-like leaves. It is stunning when covered in masses of small yellow honey-scented flowers that attract bees and other insects.

PROPAGATION
Seed or cuttings – the latter will grow faster and produce flowers sooner. You will probably find seedlings in your garden, which can be lifted and nurtured until they are large enough to plant out.

CULTIVATION
Soil type is not important, but it must be light and well-drained, and contain plenty of compost. Mulch well and replenish regularly. Feed from time to time (growing season) with small amounts of slow-release 3:1:5 fertiliser. Use this dense bushy plant in a shrub border, or mass plant it in a large flowerbed to form a groundcover. Set the young plants out close together in large groups, always in a sunny position. Site it against a west wall where it can enjoy abundant sunshine and give you a superb display of colour in return. It can even be shaped into a pretty small multi-stemmed tree. Position a large group as background shrubs, which once a year will surprise everyone with a burst of bright colour. This *Euryops* is perfect for new gardens, where it rapidly fills empty spaces. Use it as a short-term screen until slower shrubs have grown and established themselves. Consider it temporary, and replace every 4–5 years. Prune back hard after flowering, otherwise it will become very untidy. The Honey Euryops can be trimmed into a neat small tree as shown in the photograph on the right. Although it originates in the southern Cape it is quite adaptable and grows well in many parts of the country, including Pretoria.

NATURAL DISTRIBUTION
Riverbanks, hillsides and forest margins.

1,5 m × 1,5 m

Jul–Sept

Cape Gold ■ Geelsewejaartjie, Phefo-ea-loti (S.So), Impepho (Sw)

Helichrysum splendidum　　　　　　　　　　　　　　　　　　　　　　ASTERACEAE

Description and uses
Cape Gold is a fast-growing, dense, erect shrub with attractive silver-grey foliage. In summer it produces small golden-yellow flowers in flattish heads. Helichrysums are popular traditional medicine plants and this species has been used to treat rheumatism. In certain areas it is used as fuel. Some helichrysums are used as bedding material because they repel insects and parasites. Dried flowers can be used for potpourri.

Propagation
Cuttings or seed – cuttings will be easier.

Cultivation
Set this aromatic plant out in groups of 3–5, or mass plant it in a sunny position to form a superb and dazzling groundcover especially when in full flower. Plants in the shade tend to lose their beautiful silver-grey appearance and become long, lanky and a little untidy, so it is better to always place this plant in the sun. Perfect for a new garden where it will provide pretty colour while slower shrubs are growing and establishing themselves. The soil must be light, well drained and contain plenty of compost. Water moderately and cut the plant back after flowering. Take care not to overwater helichrysums in the garden especially in winter – they tend to become infected with fungus and die back in patches. Regard this plant as temporary and replace every 3–4 years.

Natural distribution
Rocky places and on forest margins.

Related species
Helichrysum patulum (Honey Everlasting, Kooigoed), which is cultivated in much the same way as *H. splendidum*, is an excellent alternative. It is a bushy shrub (1 m x 1 m) with soft felty aromatic leaves and creamy-white flowers (Sept–Feb). Once used as bedding, hence the name Kooigoed.

H. patulum

 Oct–Jan

1,5 m × 1,5 m

Ribbon Bush ■ Seeroogblommetjie, Lintbos, uHladlwana olukhulu (Z), uHlonyane (Z)

Hypoestes aristata　　　　　　　　　　　　　　　　　　　　　　　　　　　ACANTHACEAE

Description and uses
The fast-growing Ribbon Bush – so-called because the petals curl like florist's ribbon – has soft hairy dark-green leaves and pretty spikes of tubular mauve or pink flowers. The nicest thing about this plant is that it flowers just before winter when most other plants are past their prime, and the aloes aren't yet in full glory! Bees, flies and other small insects visit the flowers, providing a source of food for insectivorous birds like the Puffback, Southern Boubou, robins, thrushes and barbets. In some areas this plant is eaten as a type of spinach. Crushed leaves are traditionally used as a poultice for sore eyes hence the Afrikaans name. The Swahili use it as a remedy for chest diseases. *Hypoestes* 'Purple Haze' is a Kirstenbosch Selection – it has a more compact habit and neater dark-green foliage. Flowers are also larger.

Propagation
Sow seed in August, or propagate from cuttings. It tends to seed itself rather freely in the garden – lift the required number of plants and transplant them into containers until they are large enough to be transferred to the garden.

Cultivation
Position in dappled shade, under trees in fertile light well-drained soil that contains plenty of compost and other organic material. Apply a thick mulch layer and replenish regularly. Feed *Hypoestes* from time to time (growing season) with small amounts of slow-release 3:1:5 fertiliser. Mass plant it under large trees where it can form an effective and reasonably hardy groundcover. Perfect as a filler in an informal shrub border. Suitable for the shady south side of the house where it will provide a pretty splash of soft colour in autumn. It is an easy-to-care-for plant that generally gives little

trouble, once established. Water fairly well in summer but less in winter, and prune back hard after flowering. Use the prunings as mulch.

Natural distribution
Dry thicket, forest and damp places, from the Eastern Cape in the south to tropical Africa.

1,5 m × 1 m

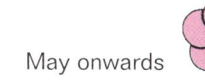

May onwards

Nodding Pincushion ■ Speldekussing, Bobbejaanklou, Luisiesboom

Leucospermum cordifolium PROTEACEAE

Description and uses
A neat symmetrical bush decorated by masses of gorgeous salmon-pink to apricot 'pincushion' flowers that make excellent cutflowers lasting for up to four weeks. If you like cutflowers for the house, make this plant your next challenge! Pincushions also attract sunbirds to the garden, which is an added bonus.

Propagation
Propagate it from seed, and protect young plants from frost. Nursery-bought plants transplant more readily when they are small (or younger) than when larger.

Cultivation
In summer rainfall areas, particularly if the soil is heavy clay, situate the Nodding Pincushion on a sloping bank or in a raised rockery for better drainage. These plants prefer acid well-drained soil and a sunny airy position. Add plenty of compost or peat to ensure that the soil is light and well drained. Water regularly, summer and winter. They grow moderately fast, and will not succeed in heavy clay soil unless it has been lightened by adding large quantities of river sand and compost or peat. As with other plants in the garden, apply a thick mulch layer (use acidic pine needles, pine bark or compost), but do not apply artificial fertiliser or manure or disturb the roots by digging. If you live on a rocky hillside or koppie, drainage will not be a problem and the plant will probably grow very well – experiment. Perfect for gardens in the south-western Cape, where it can be used as a filler in an informal shrub border, or anywhere else in the garden. If you do not plan to harvest the seed for propagation purposes, cut off the dead flowerheads after flowering. The bush can be trimmed back a little to neaten if necessary. The lifespan of the individual plants is seldom more than 10 years, if that.

Natural distribution
South-western Cape.

> *HINT: There is no limit to the amount of compost or other organic material that can be added to flowerbeds. It conditions the soil and replaces lost nutrients.*

Jul–Dec

1,5 m × 2 m

Eight Day Healing Bush ■ Agtdaegeneesbos, Douwurmbos, Lobos, Luibossie

Lobostemon fruticosus

BORAGINACEAE

Description and uses
Beautiful delicate pink or blue flowers grace this small bushy shrub in early summer, making it a real favourite with gardeners. Narrow oval hairy leaves grow all the way up soft-wooded stems that are topped by pretty bell-shaped flowers. This is another of our popular traditional medicinal plants – *Lobostemon* leaves fried in sweet oil, pulped leaves and leaf decoctions are all old Cape remedies for ringworm, sores, ulcers, burns, eczema and wounds. As the common name suggests, it is supposed to be able to heal all conditions within 8 days!

Propagation
Seeds itself freely in Cape gardens; fresh seed sown in March takes about 5 weeks to germinate. Can also be propagated from cuttings. Seedlings and young plants must be protected from frost.

Cultivation
Prefers well-drained compost-enriched soil with a thick layer of mulch that must be replenished regularly. Water throughout the year, especially in the summer rainfall region. Don't let the plants dry out during the hot, dry and sometimes windy spring months. At the coast it is fairly drought resistant. On the Highveld it likes some shade. This plant grows moderately fast and is suitable for a rockery or mixed shrub border. Grow it against the sunny north or west wall of the house for some lovely soft colour. Try planting it in a large container. Add plenty of compost to the soil mix. Use it as a temporary filler to 'pretty up' new gardens until slower shrubs have grown and established themselves. Older plants become leggy and should be discarded when past their prime. Replace with some of the younger, self-seeded plants.

Natural distribution
Common in the south-western Cape.

1 m × 1 m

Aug–Nov

Sunbird Bush ▪ Sonbekkiebos

Metarungia longistrobus ACANTHACEAE

Description and uses
Strange 'ribbon-like' yellow-brown petals are displayed in spikes above the glossy dark-green foliage of the Sunbird Bush. Bees probe busily and deeply into the unusual flowers. Black and Whitebellied Sunbirds visit them regularly too, hence the name.

Propagation
Easily propagated from seed and cuttings. Seeds are released from dry fruits with an audible 'pop'. The Sunbird Bush often self-seeds in suitable areas.

Cultivation
Mass plant this bushy rounded shrub in light shade under large trees to form an attractive groundcover, or use it on the south side of the house to pretty up the entrance. Experiment with it in a large tub on a not too shady patio – don't forget to water and feed the container regularly. Feed with slow-release 3:1:5 fertiliser at 6-weekly intervals (growing season), or use an organic plant food like Seagro, Kelpak or Nitrosol. If preferred one can even alternate the fertilisers – this sometimes helps to make plants grow better. Position *Metarungia* where it can be the centre of attraction in a townhouse garden, or in an informal shrub border. It grows fairly well in light partial shade, but seems to perform just that bit better if it receives some sun (preferably morning sun) for a couple of hours every day. It grows fast and easily (about 1 m per year) provided it is planted in fertile light well-drained soil that contains plenty of compost and other organic material. Mulch well and water regularly. Trim back once in a while to keep it neat, though it shouldn't require too much attention. Use the prunings as mulch. It will tolerate fairly hard pruning, if necessary.

Natural distribution
Rocky wooded mountain slopes.

> **HINT**: *When planting shrubs in pots or tubs, always add bonemeal (root food) to the potting mixture, which should be well drained. To ensure good drainage, add lots of coarse light compost (at least half by volume) to the mix. Put a layer of stones at the bottom of the container to make certain that the water drains freely.*

Jan–Mar, Jul–Sept

1,2 m × 1 m

Tortoise Berry ■ Skilpadbessie, Bokbessie, Cargoe, Mmaba (Tsw)
Nylandtia spinosa POLYGALACEAE

Description and uses
The Tortoise Berry is erect and stiffly branched, with small needle-like leaves. Masses of dainty mauve sweetpea-like flowers are followed by shiny scarlet berries that are edible. The slender arching stems end in a sharp spine. This is a wonderful plant for your 'people-and-bird garden' – children and fruit-eating birds thoroughly enjoy the astringent thirst-quenching berries. It is sometimes planted to control sand at the coast. Leaves and stems are traditionally mixed with those of *Lebeckia multiflora* and taken for colds and flu. A leaf and stem infusion is taken as a general tonic and for tuberculosis. The Tswana use the root to prepare a remedy for malaria. The fruits are rich in vitamin C and were collected and eaten by sailors at the Cape, to prevent scurvy (before Van Riebeeck's time).

Propagation
Fairly easily propagated from fresh seed.

Cultivation
Plant it in a shrub border, preferably situated so that the lovely flowers and fruits (and birds!) can be seen and appreciated from the house or patio. Add plenty of compost and water moderately throughout the year. The coarser and lighter the compost, the better the drainage will be. Remember that this plant originates in the drier areas, so runoff must be good. It won't be happy standing with its feet in water. In the summer rainfall region it should succeed best on a very well-drained rockery in sandy soil. Remember to water it a little in autumn and winter as well. Mulch well and replenish regularly. This spiny shrub grows moderately fast, and tends to fare better in drier areas where it is more at home. Prune to neaten if necessary.

Natural distribution
Dry areas of the north- and south-western Cape and along the coast.

1,2 m × 1,2 m

 to

Jul–Aug

Shell Bush ■ Pink Sage, Pienksalie
Orthosiphon labiatus　　　　　　　　　　　　　　　　　　　　　　　　　　　　　LAMIACEAE

Description and uses
The Shell Bush is a fast-growing shrub with soft hairy heart-shaped leaves and pretty spikes of two-lipped pale pinkish-mauve flowers.

Propagation
Easily propagated from seed or from cuttings. Sow seed (depth 1,5 x seed size) spring to summer. Use sand, bark and compost in equal quantities. Place in light shade and keep moist. Germination time, 3 weeks. Take semi-ripe hardwood, softwood or herbaceous cuttings in spring. Treat with Seradix 1 or 2. Root in equal quantities of bark and polystyrene. Place in a mist house. Rooting time, 3 weeks. Hardening off, 2 weeks. Success rate, 80%.

Cultivation
This plant is semi-deciduous, losing some of its aromatic leaves in winter, and is perfect for an informal shrub border. Mass plant it under large trees to form an effective and attractive groundcover. Provide rich well-drained soil, adding plenty of compost, and water well in summer. Plant the Shell Bush where it will receive some morning sun and shade for the hottest part of the day. This shrub in particular profits by the addition of plenty of compost, humus or other organic material, to provide it with a nice lightweight growing medium for its fine roots. It will also benefit from the addition of a thick layer of organic mulch, e.g. dried leaves, bark, compost, chopped up prunings and grass clippings (not with roots – they will grow!). Feed the plants with slow-release 3:1:5 fertiliser at intervals of 6 weeks throughout summer. Cut back hard at the end of summer/winter and replenish the mulch layer – it will quickly regrow in spring. Suitable as well for coastal gardens. Prefers temperatures between 5 °C and 30 °C and has a life expectancy of around 10 years.

Natural distribution
Rocky koppies from KwaZulu-Natal to Zimbabwe.

> **HINT**: *In general, when preparing flowerbeds, apart from 'loads' of compost and other organic material, always dig in bonemeal and/or superphosphate, to feed the roots. These can be applied at the rate of about 250 ml per square metre. Dig the bed over well and water it – leave to settle for a couple of days before setting out new plants.*
> *A sun-warmed hosepipe is perfect for marking outlines of the bed. Gentle sweeping curves are better than complicated zig-zags and contortions.*

Nov–Apr

1,5 m × 1,5 m

Heart-leaved Pelargonium

Pelargonium cordifolium GERANIACEAE

Description and uses
The Heart-leaved Pelargonium is a robust well-branched bushy shrub, usually about 1 m tall, but can reach 1,2 m × 1,2 m under favourable conditions. The heart-shaped aromatic leaves are deep green and range from smooth to hairy. Most attractive in full bloom, when pink to purple flowers with dark markings are displayed well above the handsome leaves. Flowers appear from June to January with a peak in spring. They attract many pollinating insects such as bees.

Propagation
Cuttings root freely – see *P. graveolens* (page 56) for detailed instructions. Seed of summer rainfall pelargoniums can be sown in spring. That of winter rainfall species must be sown in autumn. Sow fresh seed in a mixture of sandy loam, coarse river sand and coarse compost (equal parts). Add some fine charcoal that has been passed through a 2–3 mm sieve. Place in a suitable container, firm down and level. Seedlings may damp off (rot and die), so it is essential that the water drains freely from the container. Before sowing, it may help to treat the soil with a fungicide to help prevent damping off. Use Kaptan (5 level teaspoons to 5 litres of water). Drench seedpans and soil thoroughly before sowing. Allow soil to dry a little. Sow seed and press down well. Cover with a layer (depth 1,5 × seed size) of fine river sand and peat. Place trays in a light airy place and protect from rain. Germination time, 2–3 weeks. Transplant into 5 cm pots when seedlings have 3 pairs of leaves. Use a mix of sandy loam, coarse river sand and coarse compost (equal parts). Add bonemeal or hoof and horn meal. Pot on as soon as the roots fill the pots. When about 10 cm high, cut back to force them to branch out and flower. Feed regularly with an organic fertiliser which is not high nitrogen, or the plant will make leaves instead of flowers.

Cultivation
Add plenty of compost to the soil before planting out in the garden – it prefers light, fertile well-drained soil. Feed with slow-release 3:1:5 fertiliser (growing season). Position plants in groups of 3–5, either in a shrub border, on a rockery, or at the edge of a pond. Remember, it enjoys moist but not saturated conditions and will tolerate some salt spray. Use it to line pathways, or as an edging along the front of a flowerbed that contains taller shrubs. Trim back after flowering to neaten. This South African *Pelargonium* was first cultivated in England in 1774.

Natural distribution
Moist places on forest margins, hillsides and riverbanks.

1 m × 1 m

Jun–Jan

Rose-scented Pelargonium ■ Wildemalva
Pelargonium graveolens GERANIACEAE

Description and uses
The Rose-scented Pelargonium is a shrubby, bushy pelargonium with smallish pink flowers and wonderfully scented leaves – a touch of roses! The scented pelargoniums all contain essential oils. The components which give the characteristic rose scent to geranium oil are only found in trace quantities in this particular pelargonium. Geranium Bronze butterflies are attracted by these plants.

Propagation
Take tip cuttings (5–6 cm long) from healthy vigorous plants without rust or virus disease, from Feb to April, when the wood is neither too hard nor too soft. Use a clean sharp knife and cut just below the leaf node, or cut with a heel. Remove all leaves except top pair. Don't allow cuttings to wilt while working. Root in a mix of coarse river sand and perlite (4:1). Treat with Kaptan (5 level teaspoons to 5 litres water). Place cuttings in a cool shaded place and keep moist, not saturated. Rooting time, 3–4 weeks. Transplant into small (5 cm) pots, using a mix of equal parts sandy loam, coarse river sand and coarse compost. Add a little bonemeal or superphosphate to the mix. If preferred, root cuttings in the 5 cm pots. In this case, do not add any fertilisers. See *P. cordifolium* (page 55) for tips on how to propagate pelargoniums from seed.

Cultivation
Use the fast-growing Rose-scented Pelargonium to line pathways and along the front edge of a shrub border, where passers-by can bump it or crush the leaves to smell the lovely fragrance. Position it either right at the front of the border, or towards the middle with a colourful groundcover, such as *Othonna carnosa* or *Bulbine frutescens*, in front. An easily grown plant; add plenty of compost to the soil before setting them out. Mulch well. Feed with slow-release 3:1:5 fertiliser (growing season). Cut back to neaten after flowering. It will tolerate some salt spray.

Natural distribution
On mountains, in semi-shade in relatively moist habitats, where summers are very hot and winters are mild, in South Africa, Mozambique and Zimbabwe.

> **HINT:** The leaves and flowers of scented pelargoniums can be used to flavour or scent cakes, ice-creams, other desserts and even mashed potatoes! The flavour is enhanced by the cooking process. Do not use leaves that have been sprayed with insecticides. Each pelargonium has its own particular scent (ranging from lemon to cinnamon), released when the leaves are crushed, and many are used to make potpourri.

Aug–Jan

1,3 m × 1 m

 to

Petalidium ■ Bloubos, Moraithama (Tsw)

Petalidium oblongifolium

ACANTHACEAE

DESCRIPTION AND USES
A bushy rounded shrub with small neat grey leaves and pretty bluish trumpet-shaped flowers. The dainty petals are streaked purple, flower throats touched yellow. Flowerbuds and developing fruits are protected by delicate lacy but prominently veined bracts. Leaves sometimes appear to have a metallic hue. The flattish fruits ripen from green to brown and split to release the seeds. This striking plant makes quite an impact in the garden – its lovely blue-grey leaves contrast with all the greens. Try using the branches in a flower arrangement. The flowers last for some time, as I have discovered with great pleasure! This plant is said to be a good fodder plant for sheep, goats and cattle though some sources disagree!.

PROPAGATION
Seed or cuttings.

CULTIVATION
Normally fast-growing (50–60 cm per year), this is an excellent shrub for people living in dry areas. It is so hardy because it originates from really arid areas – in fact it is inclined to die back in patches if there is too much rain. It is advisable not to overwater it if you want it to look healthy and produce lots of its beautiful pale blue flowers. In moister areas, for best results, plant in a very well-drained position in full sun. A natural koppie or a rockery would be perfect. Add lots of compost to the planting area and resist the temptation to water it too much! I have had a couple of these plants in my own garden for about five years. They have done exceptionally well (I hardly water them), except for the years when we have had almost continuous rain. They then died back in large patches. I pruned away the dead bits, and when the rain stopped they recovered and prospered. Every year they flower well too. They don't mind pruning – each year I prune mine back quite hard to keep them neat. I feed them occasionally (about twice a growing season) with a small amount of bonemeal sprinkled under the shrubs and dug in lightly. A very light sprinkling of slow-release 3:1:5 fertiliser is also given at the same time. Mulch the plants well with the prunings. Experiment with this plant in a large decorative pot, but make sure the soil mix is exceptionally well drained. Never let the pot dry out completely, but also don't keep it too moist!

NATURAL DISTRIBUTION
In dry areas of the North West Province.

1 m × 1 m

Mar–Jul

Featherhead ■ Veerkoppie

Phylica plumosa   RHAMNACEAE

Description and uses
A fairly upright to rounded shrub, topped with the most beautiful soft feathery yellowish flowers! Flowerheads are made up of narrow, dense, fringed bracts with silvery hairs. The actual flowers cannot be seen; they are tiny and cinnamon-scented. Attractive needle-like foliage, covered with silky hairs, is neatly arranged up the stems. The decorative fluffy flowerheads, or 'feathery pompons' are popular for flower arrangements and are also used as dried flowers.

Propagation
Seed or cuttings. Sow fresh seed in Mar/Apr (depth 1,5 x size of seed). Use a well-drained loam seed mixture. Place in full sun and keep moist. Protect seedling trays from rodents and birds. Germination takes about 4 weeks; success rate, 50%. Take cuttings from a vigorously growing mother plant in Apr/May. Use tip cuttings from hardened-off new growth. Treat with Seradix 3 and root in a mixture of bark and polystyrene (6:1). Use bottom heat. Rooting time, 4 weeks; hardening off, 2 weeks; success rate, 50%.

Cultivation
Featherhead is fast-growing and prefers acid sandy well-drained compost-enriched soil, but will in fact grow in most soils, including loam and those that are nutrient-poor. Its roots are non-invasive, and it does not require fertiliser. While this shrub is probably best suited to winter rainfall areas, it may grow happily in other parts of the country too – be brave, experiment! Use as a general garden filler or plant it in an informal shrub border, where it will look stunning. In a rockery it really shows to advantage. Plant the Featherhead in full sun and water well in autumn and winter, less in summer. Prune after flowering to shape and neaten the plant. It prefers temperatures between 5 °C and 25 °C, and has a life expectancy of about 5 years.

Natural distribution
Lower mountain slopes in the south-western Cape.

Related species
Phylica pubescens (Featherhead) is very similar, growing to 1,2 m high, with lighter, smaller 'flowers'. This plant is particularly suited to seaside gardens where it will be fairly drought resistant in summer. Cultivate as for *P. plumosa*.

Jun–Jul

1 m × 80 cm

Large Spurflower Bush ■ Persmuishondblaar

Plectranthus ecklonii LAMIACEAE

Description and uses
The Large Spurflower is a soft-wooded shrub with big oval soft-textured dark-green leaves and beautiful tall spikes of purple tubular two-lipped flowers. Pink and white varieties are also available. Larvae of the Garden Inspector, Brilliant Blue and Dryleaf butterfly feed on *Plectranthus* species. A traditional remedy prepared from the leaves is used to treat headaches and hayfever.

Propagation
Very easily propagated from cuttings.

Cultivation
Mass plant this fast-growing shrub in the shade under trees to form an attractive groundcover. Perfect for the shady south side of the house, or garden wall. A single strategically placed specimen in a small townhouse garden can be the centre of attraction for a couple of weeks in autumn. If it is grown in a spot that is sunny all day the leaves tend to become smaller and a little yellow. Sun for a couple of hours a day should not be a problem. In dappled shade the plant will develop large lush-looking leaves. Set this plant out in groups of 3–5 in an informal shrub border in well-drained compost-enriched soil with plenty of moisture in summer. Remember to add an extra dose of compost and leafmould for this forest-loving plant – it thrives in the rich, light humus on the forest floor. Replenish the mulch layer regularly and feed with slow-release 3:1:5 fertiliser at intervals of 6 weeks throughout the growing season. Cut the plant back by half, or more, at the end of winter. Don't throw the prunings away – use them as mulch, or 'plant' them into small pots in a well-drained mix – they root quite easily if kept moist in a shady position. Semi-deciduous in cold gardens – plant in a protected place.

Natural distribution
Forest understorey.

1,5 m × 1,5 m

     to to

Autumn

Forest Spurflower ■ Pink Fly Bush, Spoorsalie, Muishondblaar, Vlieëbos, Cabhozi (Sw)

Plectranthus fruticosus　　　　　　　　　　　　　　　　　　　　　　　　　　　　　　　LAMIACEAE

DESCRIPTION AND USES
The Forest Spurflower is a fast-growing soft-wooded shrub, with large softly textured heart-shaped leaves that are tinged purple beneath. It produces masses of attractive pyramidal spikes of pink or bluish-mauve flowers. Larvae of the Bush Bronze, Gaudy Commodore and Garden Inspector butterfly feed on this plant. Stems rubbed on windowsills repel flies. Many *Plectranthus* species have an unusual scent when the leaves are crushed. The oils that are given off are also said to chase off flies and other insects – hence the name 'Fly Bushes'. Dried flowers and leaves are used to make potpourri.

PROPAGATION
Easily propagated from cuttings, or even from seed.

CULTIVATION
Set the plants out in groups of 3–5, in partial shade under trees or on the shady side of the house in light well-drained compost-enriched soil. Remember that *Plectranthus* are forest plants that like lots of humus and leafmould for their fine roots, so always add extra compost and leafmould when planting them. Mulch well and water copiously in summer, but less in winter. Most *Plectranthus* can be grown in large containers on a shady patio – water and feed well. Fertilise with slow-release 3:1:5 at intervals of 6–8 weeks throughout the growing season. Very attractive when mass planted to form a groundcover in a shady corner. Perfect for an informal shrub border. Use it to 'pretty-up' a shady entrance area. Cut back after flowering to encourage new growth. In humid coastal areas this plant can tolerate a little sun. Many lovely cultivars are available.

NATURAL DISTRIBUTION
Forests of the mist belt.

> **HINT**: Never water plants with a hosepipe (unless the water is finely spread through a nozzle) or with a bucket. This may compact the soil needlessly, making it almost impermeable to water. It may also dislodge or wash away your mulch.

Dec–Feb

1 m × 1 m

 to

Lobster Flower ■ Blue Coleus, Knoffelsalie

Plectranthus neochilus   LAMIACEAE

Description and uses
An unusual looking *Plectranthus* with a tall thickish flowerspike that is distinctly four-sided at the tip. Carried high above the foliage, the two-lipped flowers are deep blue and purple. The bluish upper lip is small, while the lower lip is large and boat-shaped. Leaves and stems are rather succulent-like and the plant gives off a heavy smell of garlic when crushed or damaged. This well-branched soft-wooded perennial has grey-green leaves that are sticky, softly hairy and dotted with orange glands beneath. Margins are slightly toothed. Most *Plectranthus* species carry their flowers in showy spikes and are a real asset in the garden, bearing flowers in autumn when little else is flowering. This one goes one step further: it flowers for most of summer! The oils emitted when the leaves of some *Plectranthus* are crushed are said to drive away flies and other insects (see *P. fruticosus*). This plant is also reputed to purify the air.

Propagation
The Lobster Flower's branches root as they touch the ground, forming a thick and attractive groundcover. This helps to prevent soil washing away and keeps the garden neat. One can easily propagate this plant by lifting and replanting the rooted stems. Water well immediately thereafter, and whenever necessary until they have re-established themselves.

Cultivation
Always use lots of compost and dig in bonemeal before planting out. Feed with slow-release 3:1:5 fertiliser at 6–8 weekly intervals throughout the growing season. The Lobster Flower is exceptionally fast-growing and makes an excellent groundcover for warmer, drier areas. In Pretoria it is sometimes knocked back by the frost, but grows out again in spring. To encourage and maintain a neat growth habit, prune the plant back hard just before spring. Feed, mulch and water well thereafter. Pruning also encourages more flowering and helps to prolong the plant's life.

Natural distribution
Dry thicket, open rocky woodland, South Africa, Zimbabwe, Zambia and Namibia.

40 cm × 60 cm

 –   to

Sept–Apr

Stoep Jacaranda ■ Stoepjakaranda

Plectranthus saccatus LAMIACEAE

P. zuluensis

Description and uses
A fast-growing soft-wooded shrub with small roundish light-green leaves. Sprays of delicate pale purple flowers, the largest of all the South African *Plectranthus*, appear from midsummer onwards.

Propagation
Easily propagated from cuttings.

Cultivation
Mass plant it in lightly dappled shade under trees to form an excellent and attractive groundcover. Use it to line the front of an informal shrub border, or plant in groups of 3–5 in partial shade. Don't plant it where it will be in full sun all day – it will tolerate a couple of hours of early morning sun. It also grows very well in a container indoors, next to a north-facing window, just out of direct sun. Turn the plant regularly to keep a 'balanced' shape. Use a well-drained porous potting mix that contains at least 2 parts compost to 1 part loam. Add bonemeal to feed the roots. Feed every 6–8 weeks in growing season with a balanced organic fertiliser. Prune now and then to neaten and encourage flowering. Repot in new soil in late winter. Plant in a protected position in cold gardens. *Plectranthus* prefers very light well-drained soil that contains plenty of compost. It also appreciates and benefits from a good layer of organic mulch that is replenished regularly. Water well in summer while it is actively growing, less in winter. Prune back reasonably hard at the end of winter, to neaten and promote vigorous growth and prolific flowering.

Natural distribution
In semi-shade in forests and rocky places.

Related species
Plectranthus zuluensis (Zulu Spur Flower, Zoeloe-muishondblaar) also grows fast, and produces spikes of pale blue-mauve flowers throughout the year. This soft-wooded shrub (1,5 m × 1,5 m) is suitable for containers and shady spots in the garden. Treat as for other *Plectranthus* species.

P. zuluensis

80 cm × 80 cm

Midsummer onwards

Ash Bush ■ Wild Rosemary, Asbossie, Laventelbossie, Vaalbossie, Wilderosmarien, T'kaibebos (Nama)

Pteronia incana ASTERACEAE

Description and uses
Attractive silver-grey aromatic foliage and masses of yellow daisy flowers distinguish the T'kaibebos. Rounded and bushy, it has a rather tangled appearance, but in full flower is really beautiful. Leaves are tiny, almost needle-like and have a very strong 'herby' smell when crushed – another plant that takes one straight to Namaqualand! The pale grey foliage contrasts well with other green shrubs in the garden. It appears that this plant is sometimes palatable to animals, but at other times not. The leaves contain essential oils occasionally used for perfume.

Propagation
Easily propagated from cuttings or seed collected in October. Sow seed (March/April) in a very well-drained mix (depth 1,5 x seed size). Keep moist. Seed germinates rapidly. Prick out in September and transplant into a well-drained loamy mix.

Cultivation
This plant originates in the drier parts of the country and does not like to be overwatered. Fairly fast-growing, it is an excellent choice for water-wise gardens. Perfect for small to large gardens in drier areas – it would probably be happy anywhere in full sun. Plant it in informal shrub borders or against a sunny west wall. Use it to line a pathway, but remember to allow enough room for it to spread. Passers-by can brush into it and enjoy the lovely herby fragrance. In moister areas, try growing it on slopes and banks. Large rockeries or a natural koppie would be good places for it to show itself off to advantage. Plant the Ash Bush in a very well-drained position and add some compost when planting. Apply a thick mulch layer and water carefully until the plant is well established. Prune if necessary to neaten. This is probably best done after flowering, unless you want to harvest the seeds for propagation purposes.

Natural distribution
In dry rocky places; also other dry habitats.

1 m × 1 m

Aug–Sept

Brown Salvia ■ Beach Salvia, Golden Salvia, Strandsalie, Sandsalie, Bruinsalie, Geelblomsalie

Salvia africana-lutea

LAMIACEAE

Description and uses
Brown Salvia is a sturdy rounded shrub with attractive grey foliage. The strangely coloured yellowish flowers fade to brick-red and then reddish-brown. They are filled with nectar and attract sunbirds. Aromatic soft grey leaves are oval in shape, and are often used as a substitute for the herb sage when cooking meat and fish. This striking shrub is traditionally used to treat coughs, colds, fevers and sometimes women's ailments. Once widespread, it is now becoming scarce as urban development destroys its habitat. Larvae of the blues and bronzes (butterflies) feed on salvias.

Propagation
Easily propagated from seed or cuttings. Seed sown in spring germinates well and seedlings grow fast. Prick them out into pots or plastic bags and plant out into good, well-drained, well-composted garden soil when 15–20 cm high.

Cultivation
A fast-growing plant; it should flower within 12–18 months of sowing. Water Brown Salvia well in the growing season (winter). In summer at the coast it can withstand drought. In cold places, position it in a warm, sheltered spot, otherwise it will be cut down by frost. Feed every couple of months with a little slow-release 3:1:5 fertiliser (growing season). Prune from time to time to neaten, particularly after the flowering season. Salvias are wonderfully easy to grow and many are tolerant of sea winds. Their attractive foliage is often grey to grey-green and aromatic year-round. The distinctive two-lipped flowers are produced in profusion. Besides sunbirds, they attract bees and butterflies. These plants grow easily in any light, well-drained soil, are happiest in full sun and once established are fairly drought resistant. For a better crop of flowers, water in the months directly preceding the flowering season.

Related species
Salvia chamelaeagnea (Aromatic Sage, Blue Salvia, Blousalie, Bloublom, Afrikaanse-salie) is a fairly fast-growing, bushy, sprawling shrub (1 m × 1 m). Reasonably frost hardy, it carries tall spikes of attractive pale violet-blue flowers in summer. Leaves are very strongly scented. Traditionally used to treat coughs, whooping cough, colds, bronchitis and diarrhoea. An exciting new *Salvia* recently discovered in the eastern Cape by Ernst van Jaarsveld, **Salvia thermara**, shows great promise as a garden subject. It has deep-red flowers, grows to about 60 cm high and is easily grown from cuttings. Kirstenbosch is propagating it at present. Another unusual and pretty *Salvia* is **Salvia muirii**, which has relatively large deep-blue flowers that are flecked white on the lower lip. The grey-green leaves smell of 'Vicks'. It reaches 30 cm and flowers from Jan to May.

Natural distribution
Dunes of the southern Cape coast.

 Early Aug–Nov

1,5 m × 1,5 m

Blue Lips ■ Blou-lippe, Mazabuka (Sw), isiThibothi (X)

Sclerochiton harveyanus   ACANTHACEAE

Description and uses
Blue Lips usually forms a roundish to spreading soft-wooded shrub with deep-green foliage. It may on occasion develop into a small tree of up to 4 m in height; it may also scramble into surrounding vegetation. Tiny strangely shaped blue-mauve to purple flowers that seem to have lips are followed by beaked capsules which split into 5 valves.

Propagation
Take semi-ripe hardwood cuttings in spring. Treat with Seradix 2 and root in a mix of bark and polystyrene (equal quantities). Place in a misthouse. Rooting time, 3 weeks; hardening off, 2 weeks; success rate, 40%. Plant it in fertile well-drained soil (preferably neutral to acid loam) that contains plenty of compost and leafmould.

Cultivation
Remember that this fast-growing shrub is a forest-loving creature, used to having its feet in a thick layer of leafmould, so for the best results always give it an extra dose of compost and/or other organic material. Mulch well and water regularly in summer. One of the advantages of this shrub is that it tolerates a fair amount of shade, and happily grows under large trees, behind big shrubs, or alongside shady walls. It sends out long shoots or runners that are inclined to root when they touch the ground, but not to the extent that the plant becomes a nuisance. In fact, if you want to cover a large area, this is a very useful attribute. Use it in an informal shrub border that receives a fair amount of shade, or plant it in a large container on a shady patio where the small strange flowers can be appreciated at close range. Position in a shady south entrance, or along a pathway in the forest area of the 'bird garden'. Suitable for coastal gardens. Prune at the end of winter to neaten. Prefers temperatures from 5 °C to 30 °C, and has a life expectancy of about 10 years.

Natural distribution
Margins of evergreen forest or as part of the forest understorey, from the Eastern Cape to Zimbabwe. NTN 681.4

1,5 m × 3 m

 to    to

Dec–Mar

Crane Flower ■ Bird-of-Paradise, Kraanvoëlblom, Geelpiesang, isiGude (Z)
Strelitzia reginae　　　　　　　　　　　　　　　　　　　　　　　　　　　　STRELITZIACEAE

Description and uses
This striking perennial is one of South Africa's most successful exports! It forms sturdy clumps of grey-green banana-like leaves and boasts unique and wonderful bird-like orange-and-blue flowers (excellent cutflowers). These produce abundant nectar that lures insects and birds such as Whitebellied, Black, Grey, Collared, Malachite and Marico Sunbirds, Cape White-eye and weavers. A variety of birds eat the seeds. A traditional remedy prepared from the flower is used to deal with swollen glands.

Propagation
Propagate it from seed.

Cultivation
Young plants are slow to establish, but if conditions are suitable, they may flower freely after about 2 years. Always plant it in good rich fertile soil, add lots of compost and water regularly. Apply a thick layer of mulch and replenish regularly. Feed with slow-release 3:1:5 fertiliser at intervals of 6–8 weeks throughout the growing season. It flowers well only when properly established, so do not divide or transplant unnecessarily. It also produces more flowers if placed in full sun. Both beautiful and useful, this fairly slow-growing wind-resistant shrub fares well in coastal gardens. Popular with landscapers in South Africa and abroad, this architecturally pleasing plant is ideal for modern landscapes, creating an impact not only in home gardens, but also in office complex gardens, schools and large parks. Perfect for hotel grounds and game park reception areas. This *Strelitzia* can form an impressive groundcover when mass-planted in very light partial shade. Use it in huge containers or planter boxes to decorate a patio or front door. Create a stunning focal point by grouping 3–5 of these plants alongside a water feature (depending on its size). In frosty areas position the Crane Flower against a north- or west-facing wall. Cut off any damaged or dying leaves. If you do not want to collect the seed for any purpose, cut the old flowerstalks back to ground level after flowering.

Natural distribution
Rocky grassland.

1,2 m × 1,2 m

 Mar–Oct

Cancer Bush ■ Turkey Flower, Kankerbossie, Belbos, Gansies, Jantjie-Bêrend, Kalkoentjiebos

Sutherlandia frutescens   FABACEAE

DESCRIPTION AND USES
Pretty silver-grey foliage, bright orange-red sweetpea-shaped flowers and attractive balloon-like pink-red pods characterise the short-lived Cancer Bush. Stock browse the foliage, sunbirds pollinate the flowers, and the caterpillars of the Lucerne Blue butterfly feed on *Sutherlandia*. The Cancer Bush is an old Cape remedy for stomach problems and cancer (as a preventative and as a treatment). It is one of the most multi-purpose and useful of South African medicinal plants. Traditionally it is believed capable of curing colds, flu, asthma, chicken pox, heart failure, diabetes, backache and rheumatism. The Khoikhoi and Nama people used decoctions to wash wounds, and drank the mixture for fevers and other ailments. Research suggests that there is indeed a scientific basis to back the traditional usage of this plant to treat many serious medical conditions, including possibly pancreatic and other cancers.

PROPAGATION
This shrub seeds itself freely, so older plants can be removed when past their best.

CULTIVATION
Fast-growing and easy-to-grow, it tolerates all soil types, and should be watered well in winter, especially in the summer rainfall areas. Add plenty of compost to the planting area. Set plants out close together in groups of 5 or more on a rockery or in a mixed shrub border. Plant a few in a large pot to temporarily decorate a patio. Use as a provisional filler in a new garden while permanent plants are establishing. Useful for lining a pathway, or grown against the sunny west wall of the house for some striking colour. Feed with a little slow-release 3:1:5 from time to time (growing season). Plants tend to become a little untidy after flowering; trim them back to neaten. Remember to collect the seed if you want to have a continual supply of robust young plants to replace the older ones in your garden.

NATURAL DISTRIBUTION
Rocky, sandy areas in South Africa, Botswana and Namibia.

1,25 m × 1 m

Jul–Dec

Krantz Aloe ■ Kransaalwyn, Sekgopha (N.So), uNomaweni (X), inKalane encane (Z), umHlabana (Z)

Aloe arborescens   ASPHODELACEAE

DESCRIPTION AND USES
The large succulent blue-grey leaves of this rounded shrub carry beautiful spikes of brilliant-red tubular flowers high above them. These produce abundant nectar that attracts not only sunbirds, but also the Crested Barbet, Cape White-eye, Blackheaded Oriole, Streaky-headed Canary, Yellowthroated Sparrow, Blackeyed Bulbul and Grey Lourie. All these birds visit the *A. arborescens* in our garden, and some of them even eat the flowers! Wood hoopoes probe with long red beaks for insects under dead, dried leaves. Fresh leaf sap is used to treat bruises, burns, abrasions and other skin complaints, and has also been used to deal with X-ray burns.

PROPAGATION
Root stem-end cuttings in coarse river sand and keep fairly dry to prevent rot. It is a long and fairly difficult process to propagate aloes from seed. Seed takes 3–4 weeks to germinate and seedlings require adequate moisture and protection from frost while still young. The plant will flower three to five years after the seed was sown. For more detailed instructions on propagating aloes from seed see *A. tenuior* (page 34).

CULTIVATION
Aloes are very special garden plants – in addition to nectar, they provide striking colour in a dull and tired-looking winter garden. The Krantz Aloe makes an excellent focal point on a rockery and is suitable for gardens both large and small. It has been used as a hedge to protect stock and crops and can be situated on a bank to help control erosion. Fast-growing and virtually problem-free, this aloe needs well-drained compost-enriched soil. It tolerates a fair amount of neglect, adapts well to both the summer and winter rainfall areas and even takes some salt spray. Some neglect is fine, but aloes are actually hungry feeders and benefit greatly from a well-prepared soil that contains sufficient nutrients and humus. Prepare the beds well before planting. Apart from compost, preferably rough and light, and other organic materials, one can add wood ash, bonemeal, river sand to improve drainage, and very well-rotted manure (not chicken, it may burn the roots). Dig in thoroughly. Water moderately in summer. Good drainage is of the utmost importance – aloes usually grow on slopes where runoff is good. Feed from time to time in the rainy season with a sprinkling of slow-release 3:1:5 fertiliser.

NATURAL DISTRIBUTION
High mountainous areas in South Africa, Mozambique, Zimbabwe and Malawi. NTN 28.1

> **HINT:** *In the garden, it is a good idea to group aloes that originate in the same area and have similar soil and watering requirements (e.g. alkaline soil and summer rain). White scale can sometimes be a problem, often because the aloe is planted in the wrong position, e.g. too much shade.*

 May–Jul

2 m × 3 m

Bitter Aloe ■ Bitteraalwyn, Tapaalwyn, umHlaba (Z, X, So), iKhala (X)

Aloe ferox   ASPHODELACEAE

DESCRIPTION AND USES
This attractive single-stemmed aloe has thick rosettes of thorny succulent leaves and tall stunning spikes of tubular orange-red flowers that are carried on showy candelabra-like flowerheads. The cut leaf exudes copious thick yellow juice which is collected and concentrated by boiling to a dark brown lumpy substance called Cape Aloes. This is used to produce a purgative drug and has been exported for more than 200 years. It is used in several medicines, e.g. 'Lewensessens'. On farms, leaf gel cut into blocks is used to make 'konfyt' (preserve), and peeled leaves make a good jam. The thickish outer layer of skin is first peeled from the leaves. The fleshy inner portion is sliced, cut up and pricked before being soaked in lime water. It is then boiled to make jam. This jam apparently has no bitter taste but I've not tasted it myself, so can't verify this! The gel is also used in hair and skin care products. Split or crushed leaves are applied directly onto sunburn, open wounds, sores, burns and scalds, ulcers and itchy insect bites (fleas or mosquitoes). Fresh juice is used for inflammation of the eyes. Dried, ground leaves are used as snuff.

PROPAGATION
Propagate it from seed (pre-treat with fungicide, Apron C) and protect the young plants from frost. For more detailed instructions on propagating aloes from seed see *A. tenuior* (page 34).

CULTIVATION
The slow-growing Bitter Aloe provides an excellent focal point for a rockery and must be planted in well-drained and compost-enriched soil. Ideal for gardens both small and large. Water moderately in summer. Remove dry dead lower leaves to keep the plant neat. The root system is non-aggressive. It tolerates salt spray.

NATURAL DISTRIBUTION
Bush scrub, hillsides and rocky mountain slopes. NTN 29.2

RELATED SPECIES
Similar in shape and size is **Aloe marlothii** (Mountain Aloe, Bergaalwyn, UmHlaba [Z], Mogopa [Tsw]), which bears attractive horizontal spikes of tubular orange or red flowers from May to September. Rich nectar draws Blackheaded Oriole, barbets and sunbirds. Widespread on mountain slopes from KwaZulu-Natal northwards to Mpumalanga. Cultivate as for *A. ferox*. NTN 29.5

2–5 m × 1 m

May–Sept

Wild Bush Petunia ■ Green's Barleria

Barleria greenii

ACANTHACEAE

DESCRIPTION AND USES
Spectacular azalea-like flowers cover this round bushy shrub in autumn. The large delicate flowers range in colour from pure white to pale pink and dark pink with magenta streaks. These plants flower prolifically and are beautiful garden subjects. Their stiff, thick-textured dark-green leaves are fairly glossy above and yellow-green below. Flower bracts and stems are a little spiny (thorny bracteoles). The lovely flowers, actually the mature unopened flowerbuds, have a strong sweet fragrance at night and attract hawk moths, which feed on them at night. It is unlikely, however, that they are the pollinators – they remove nectar via a slit on the side of the flower! The flowers are filled with abundant sweet nectar and attract bumblebees. An asset to any garden, this plant somehow managed to keep itself well hidden until 1984, when Dave Green, the new owner of a farm near Estcourt, Natal, discovered it on his property! Fortunately for gardeners, he brought it to the attention of botanists, and it is now becoming available at various nurseries.

PROPAGATION
Seed and cuttings. Germination rate may be about 60%. Seedlings grow rapidly and can be planted out in well-drained soil when about 20 cm high. Add plenty of compost and mulch well. Water young plants thoroughly directly after planting, and thereafter carefully until they are properly established. After that, don't overwater. They will probably flower when they are about 18 months old. Use bottom heat for the cuttings.

CULTIVATION
Perfect for gardens both small and large. This plant tolerates a wide range of climatic and soil (acid or alkaline) conditions, both inland and at the coast – so every garden should have at least a few! For a stunning display plant them in groups of 3–5 in an informal shrub border or along a pathway. Remember, they're a little spiny, so keep them well away from the path itself. Experiment with them on a rocky koppie or plant amongst scattered thorn trees. This *Barleria* seems to produce more flowers when there is less rain, so it would be advisable not to overwater it. Green's Barleria is quite unfazed by frost and some snow, so if you have a really cold garden, this may be just the plant to experiment with. Prune in late winter or early spring to neaten, to encourage dense bushy growth and prolific flowering. I examined the plants growing in the Pretoria National Botanical Garden in June. They had dropped their leaves, but did not appear to have any frost damage. Possibly in warmer areas the plant is evergreen?

NATURAL DISTRIBUTION
On heavy black clay, between dolerite boulders on north-facing slopes.

1,8 m × 1,2 m

Feb–Mar

Medium shrubs 71

Pride-of-De Kaap ■ Lowveld Bauhinia, Vlam-van-die-Vlakte, Beesklouklimop, Motshiwiriri (N.So), Tswiriri (Sha), Kisololo (Sw), umVangatane (Z)

Bauhinia galpinii FABACEAE

DESCRIPTION AND USES
Pride-of-De-Kaap is a magnificent rambling and spreading shrub that has two-lobed leathery light-green leaves and bright salmon-pink to brick-red orchid-like flowers. The flowers attract hordes of butterflies: larvae of the Brown Playboy and Orange Barred Playboy butterfly feed on the seeds in the pods, while larvae of at least three other butterfly species (e.g. Koppie and Foxy Charaxes) eat the leaves. Game browse the leaves, and the flower-buds and green seeds are eaten by Grey and Purplecrested Louries.

PROPAGATION
Soak seed overnight in warm water (not hotter than 50 °C). Sow spring to summer, 5 mm deep, in sandy well-drained loam. Place tray in light shade and keep moist. Feed seedlings with Seagro/Kelpak after germination. Success rate, 80%. Protect young plants from frost.

CULTIVATION
In full flower Pride-of-De Kaap is one of our most beautiful shrubs and is suitable for lining spacious driveways, for an informal hedge, a mixed border or a rockery. Ideal for medium to large gardens. Fairly hardy and fast-growing, it has been successfully planted in the centre of highways and is perfect for large office complexes, school grounds and parks. It grows well in any type of soil, but always add plenty of compost and/or old kraal manure to the planting area for the best results. Feed with slow-release 3:1:5 fertiliser at intervals of 6–8 weeks throughout summer. Evergreen under favourable conditions and fairly drought resistant in winter, it does not require much water at that stage. It must be well watered in summer. This *Bauhinia* does tend to scramble and spread into other shrubs – prune to neaten if necessary. Tolerates temperatures between –5 °C and 25 °C and has a life expectancy of around 20 years.

NATURAL DISTRIBUTION
Woodland and riverine bush, from KwaZulu-Natal to Zambia. NTN 208.2

HINT: Shrubs that tend to 'lean into' or all over the plants around them need to be controlled to some extent, otherwise they may get out of hand. Most grow fast and produce lots of flowers and/or fruits.

3 m × 5 m

 to   to

Dec–Apr

Natal Bauhinia ■ White Bauhinia, Dainty Bauhinia, Fynbeesklou, Natalsebeesklou
Bauhinia natalensis FABACEAE

Description and uses
An attractive neatly rounded ornamental shrub that is 'snowed under' for most of summer with delicate, somewhat crinkly white flowers, each of which has petals streaked deep-pink or red. Pale-yellow thinly woody pods split open in winter, releasing the brown seeds with a vigorous pop. Whitebellied Sunbirds regularly visit the lovely fragrant flowers. Caterpillars of the Common Emperor and Banded Achaea moth feed on *Bauhinia*.

Propagation
Seed germinates readily in 10–12 days. It may seed itself freely in the garden if conditions are good – give some of the extra plants to a lucky friend. Seedlings grow fast in a commercial potting mix. Transplant into the garden when they are about 15 cm tall. They may flower in their second season.

Cultivation
This dainty fast-growing shrub is ideal for smaller gardens, or it may be planted in groups of 3–5 for greater effect in larger gardens. Perfect for an informal shrub border. Remember to allow enough room for it to spread. It looks wonderful planted next to a water feature or pond, especially when the sunbirds visit. For best results plant it in the sun, although it can tolerate a little shade for part of the day, and provide it with plenty of compost and other organic material when planting. Apply a mulch of similar material and replenish regularly as it decomposes. This shrub seems to grow better and more robustly in a fertile, rich loam soil. A light sprinkling of slow-release 3:1:5 can be applied at approximately 8-weekly intervals throughout summer. Water moderately throughout the year. Naturally neat, it doesn't usually need much pruning, although a light trimming once in a while won't come amiss. The Natal Bauhinia produces more flowers when it is not overwatered.

Natural distribution
Valley bushveld and scrub. NTN 208.5

 Nov–Apr

2,5 m × 3 m

Yellowbell Bauhinia ■ Bosbeesklou, Geelbeesklou, isiThibathibana (Z, N.So)

Bauhinia tomentosa FABACEAE

Description and uses
An attractive shrub with small light-green two-lobed leathery leaves. It is free-flowering, decorated for months with beautiful yellow bell-shaped flowers, each with a black marking in the throat. These flowers attract pollinating insects, which in turn draw insectivorous birds such as the Barthroated Apalis, Cape White-eye, Crested and Blackcollared Barbets, and Natal and Cape Robins. The Grey Lourie eats the flowers. Caterpillars of the Gold Bordered Prince moth feed on it.

Propagation
Propagate it from seed. Protect the young plants from frost.

Cultivation
This *Bauhinia* grows moderately fast and is suitable for a rockery or mixed shrub border in gardens both small and large. Set plants out singly, or in groups of 3–5 in larger gardens, to create a lovely impact. Experiment with it in large decorative containers on a sunny patio or next to a pool, where the pretty flowers can be admired at close hand. It also makes a lovely bonsai, planted in a small container in well-drained soil. This plant may be evergreen in a mild climate, and prefers light well-drained soil with plenty of compost added. Mulch well and replenish the layer regularly. Feed with slow-release 3:1:5 fertiliser at 8-weekly intervals throughout summer. Prune the plant to keep it in shape.

Natural distribution
Woodland, riverine bush and coastal dune bush in South Africa, Mozambique and Zimbabwe. NTN 208.1

2 m × 3 m

 to

Dec–Mar

Giant Salvia ■ Tropical Salvia, Reusesalie

Brillantaisia subulugurica ACANTHACEAE

DESCRIPTION AND USES
The Giant Salvia is a nicely rounded shrub or perennial with large, softly textured, somewhat glossy, dark-green leaves and long spikes of big blue-mauve salvia-like flowers. Burchell's Coucal, the Southern Boubou, Cape Robin and Crested Barbet like to hunt around in the depths of this shrub. Common Mother-of-Pearl butterfly larvae feed on the leaves.

PROPAGATION
Easily propagated from cuttings – robust clean-cut stems will even root if inserted directly into the soil especially in warmer climates. However, it is usually better to plant the cuttings in clean river sand in pots.

CULTIVATION
To grow really well the Giant Salvia needs full sun for at least part of the day – too much shade limits flowering and makes it long and lanky. Its large lush leaves lend an immediate tropical air to the garden, enhanced by the strikingly large flowerheads. Suitable for gardens both small and large. Plant this vigorous fast-growing shrub in a mixed shrub border, or in planter boxes, in good light well-drained soil. Plants with lots of fine roots, like this Salvia, will perform better with plenty of added compost. Mulch well and replenish the layer regularly. Water well in summer, but less in winter. Feed with slow-release 3:1:5 fertiliser at intervals of 6–8 weeks throughout summer. The frost-tender Giant Salvia needs to be planted in a very protected position. Trim back lightly after flowering, to neaten, but cut back by at least half, if not more, after winter to keep it neat and encourage bushiness. Use the prunings as mulch, or insert directly into the soil to increase the size of the bush.

NATURAL DISTRIBUTION
Zimbabwe and tropical Africa.

> **HINT:** When inserting cuttings directly into the soil, always use a thickish stick (or screwdriver) to make the holes beforehand. Cuttings with jagged ends may rot and die.

Summer

2 m × 2 m

     to

Wild Pomegranate ■ Wildegranaat, Buffelshoring, Mahlozana (Sw), umFincane (X), iThobankomo (X), isiGolwane (Z), uQongqo (Z)

Burchellia bubalina RUBIACEAE

DESCRIPTION AND USES
A neat shrub with glossy dark-green foliage, decorated by clusters of tubular nectar-rich orange to coral-red flowers. Whitebellied, Collared, Olive and Black Sunbirds are lured by the nectar and are constant visitors to the flowers in early summer. In the shade, this plant can grow into a small tree of up to 5 m (width 2,8 m). Roots are traditionally added to body washes and used to prepare a love charm. The hard, dense close-grained wood is used to make farm tools.

PROPAGATION
Seed sown in October takes about 4–6 weeks to germinate; young plants must be protected from frost. Take semi-ripe hardwood cuttings from an actively growing mother plant in autumn or spring. Treat with Seradix 3 or IBA 6 000 p.p.m., 5-second drip. Use a mix of bark and polystyrene chips (equal volumes) with bottom heat (28 °C) and intermittent mist. Rooting takes 8–10 weeks; hardening off, 4 weeks; success, 79%. Selected plants are best propagated by cuttings.

CULTIVATION
One reason for this shrub's popularity is the fact that it starts flowering at an early age. The Wild Pomegranate is a slow-growing attractive ornamental that will tolerate partial shade, but needs to be planted in a position where it is protected from cold winter winds. It thrives where rainfall is high, and in humid coastal gardens where there is no frost. Place it in a position where it receives a little shade, especially at the hottest times of day. Plant it under large trees or in a mixed shrub border. This plant has a non-invasive root system and makes a nice hedge. It even does well in a large container on a shady patio. Remember to feed and water it well. *Burchellia* prefers rich well-drained loam soil, preferably acid, and tons of compost or peat! Mulch well and replenish regularly; water well in summer, especially if there is no rain. There is usually no need to prune it, as it tends to stay neat. Prefers temperatures from –5 °C to 25 °C and has a life expectancy of about 30 years.

NATURAL DISTRIBUTION
Forests of the east coast. NTN 688

HINT: Compost is not manure! Manure is an animal product, compost is a plant product.

2,5 m × 1,5 m

     to

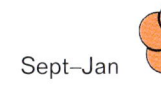

Sept–Jan

Medium shrubs

Bushtick Berry ▪ Bietou, Boetabessie, Bokbessie, iTholonja (Z), Motlempa (S.So), uLwamfithi (X)

Chrysanthemoides monilifera ASTERACEAE

DESCRIPTION AND USES
The Bushtick Berry is a large, dense and spreading soft-wooded shrub with grey-green foliage that produces lots of pretty yellow daisy flowers and sweet tasty purple, black or red berries. They were once an important food source for the Khoikhoi, but are now mainly eaten by children, who love them. Birds such as the African Green Pigeon, Fiscal Flycatcher, Redwinged and Glossy Starlings, Crested and Blackcollared Barbets, Karoo Robin and Collared Sunbird really enjoy the fruits. Larvae of the Common Opal, Beaufort Opal, Natal Opal, Coast Copper, Thysbe Copper and Jitterbug Daisy Copper butterflies feed on this plant. Many animals, such as the Blue Duiker, like to browse the leaves. Soap has been made from the ashes. Traditionally, leaf infusions are used to deal with fevers. Fruit juice is taken in small doses to purify the blood and deal with impotence.

PROPAGATION
Easily propagated from cuttings.

CULTIVATION
The fast-growing Bushtick Berry is drought- and wind-resistant and tolerant of coastal conditions. It is ideal for gardens both small and large, grows easily in any soil and makes an excellent hedge, windbreak or screen plant for the coast (will grow on dunes). To encourage better growth, always add plenty of compost and other organic material to planting holes, mulch and water regularly. In the growing season, feed with slow-release 3:1:5 fertiliser about once every 8 weeks. Plant it singly or in large groups to form an attractive and effective groundcover, even under large mature trees. Use it as a temporary filler, especially in new gardens, until slower shrubs have grown and established themselves – in the wild, this is a pioneer species and is often one of the first to colonise

disturbed areas. Prune whenever necessary to keep it neat and within bounds. This plant grows very well in the Pretoria area – experiment with it elsewhere too.

NATURAL DISTRIBUTION
South-western Cape, mainly on the flats and on mountain slopes, north to Zimbabwe. NTN 736.1

 Spring

2 m × 2,5 m

Dragon Tree ■ Large-leaved Dragon Tree, Drakeboom, Grootblaardrakeboom, Photsoloma (N.So), umKhoma-khoma (X), iTokothoko (Z), iGonsi-lasehlathini (Z)

Dracaena aletriformis (= *D. hookeriana*)　　　　　　　　　　　　　　　　DRACAENACEAE

DESCRIPTION AND USES
The Dragon Tree is a wonderful foliage plant for shady areas in slightly warmer gardens. Usually single-stemmed, it has large leathery dark-green strap-shaped leaves, and lovely tan bark marked with leaf scars. Tiny sweetly scented (at night) yellow-green flowers are carried on tall spikes, and are followed by beautiful orange berries. These are popular with many of the fruit-eating birds. The flowers, pollinated by hawk moths, attract insects with their following of insectivorous birds – this is an excellent plant for the 'bird garden'. Caterpillars of the Bush Nightfighter butterfly feed on *Dracaena*.

PROPAGATION
Propagate the Dragon Tree from seed or stem cuttings. Seed should be sown in an organic potting mix and kept moist until it has germinated (about 3 weeks). Seedlings grow reasonably fast. When 10 cm high, transplant into small individual (about 500 ml) containers. Use a mix of equal parts coarse river sand, compost, light organic potting soil and good garden soil, preferably loam. Add approximately 5 ml (1 teaspoon) each of superphosphate, bonemeal and slow-release 3:2:1 fertiliser. Water well and then allow the soil to become fairly dry before watering again. Stand each container on a layer of pebbles in a drip tray to ensure good drainage. Stem cuttings (clean cut) can be rooted in coarse river sand in individual containers. Keep moderately moist.

CULTIVATION
This attractive foliage plant needs a very protected position in colder gardens and makes an exceptional container plant for indoors or for a shady patio. It will tolerate fairly poor light conditions and must not be situated in direct sunlight, or the leaves will burn. Fairly fast-growing; plant in deep, fertile (especially clay or good loam), well-drained compost-enriched soil (don't stint on this!) and water moderately. Mulch well and replenish regularly. Feed with slow-release 3:2:1 fertiliser at 6–8 weekly intervals throughout summer. Cut off any old dry leaves.

NATURAL DISTRIBUTION
A variety of habitats – shady places in the dry bushveld, dune forest undergrowth and mountain forests. NTN 30.9

2 m × 1,5 m

Nov–Feb

Dwarf Coral Tree ■ Kleinkoraalboom, Mokhupye (N.So), umSintsane (X), iKati (Z), umSinsana (Z)

Erythrina humeana   FABACEAE

DESCRIPTION AND USES
This attention-grabbing shrub (or small tree) just demands a place in every garden – and a well-deserved place it is too. Leaves are made up of three leaflets – hook thorns are found on main veins and leafstalk. The grey-green to grey stems are also covered with scattered hook thorns. Long-lasting nectar-rich flowers – a brilliant showstopping red in full flower – are produced throughout summer and attract both insects and birds. The Cape Parrot feeds on the nectar, as do Collared, Olive and Greater Doublecollared Sunbirds. The Brownheaded Parrot eats the unripe seeds and the Grey Lourie the flowers. Traditional remedies prepared from the roots are used to treat sprains, wounds and swellings.

PROPAGATION
Easily propagated from truncheons, or seed that has been covered with boiling water and allowed to soak overnight before sowing. Sow in levelled river sand. Press seeds into sand (depth 1,5 x seed size). Cover with a thin layer of sand. Place trays in a hot area and keep moist. Germination time 2–3 weeks. Transplant into small containers when seedlings have about 2 leaves. Plant into the garden when about 1 year old. The fastest method of propagation is to plant truncheons. These strike easily if planted in August.

CULTIVATION
The Dwarf Coral Tree has a distinctive shape and the attractive blooms allow it to be used to advantage as a focal point in the garden or on a rockery. In smaller gardens that have no room for large trees, it could be pruned into a nicely shaped small tree. Alternatively, it could be added to the shrub border and allowed to develop a bushy form. In suitable climates, it is perfect for home gardens (small to large), office complexes and hotel grounds where the flowers will show off vividly and create a superb African impact. It grows fast, is reasonably drought tolerant, and thrives if plenty of compost is added to the planting area. Feed with slow-release 3:1:5 fertiliser at intervals of 6–8 weeks throughout summer. Water moderately in the growing season. Prune to neaten and shape when necessary. An excellent tree for warmer bushveld areas, where there is very little frost.

NATURAL DISTRIBUTION
Mountainsides and rock outcrops, in bushveld (may reach 4 m), dry scrub and on coastal dunes, from the Eastern Cape to Mozambique. NTN 243.1

HINT: When planting truncheons out in the garden, always place clean river sand at the bottom of the hole. This prevents rotting and encourages root formation.

 Sept–Apr   3 m × 2,5 m

Blue Honeybells ■ Inyanga Hedge Plant, Blouheuningklokkiesbos

Freylinia tropica　　　　　　　　　　　　　　　　　　　　　　　　SCROPHULARIACEAE

DESCRIPTION AND USES
Small delicate-looking blue-mauve flowers shelter amongst the shiny bright-green leaves of this rather erect, narrow and sparse shrub. Flowers may also be white or lilac. In the wild this shrub occasionally forms a small tree of 3–5 m high. It attracts lots of butterflies!

PROPAGATION
Seed or cuttings. The tiny wingless seeds germinate readily within three weeks. To propagate from cuttings use stem cuttings that are taken during the warmer summer months. If conditions are suitable, young plants may grow rapidly and flower within a couple of seasons.

CULTIVATION
Suitable for gardens both small and large. Compensate for the Blue Honeybell's narrow shape by grouping a number of plants close together (about 30 cm apart) when you set them out – this will ensure that you end up with a nice bushy effect. Mass plant them in very light shade under trees to form a groundcover, or plant them in groups of 3–5 plants in a shrub border. Use them to line a pathway or to form a bit of a screen – check that they are planted in fertile light well-drained soil with plenty of compost! Feed with slow-release 3:1:5 fertiliser at intervals of 6–8 weeks throughout summer. Plants treated in this manner do extremely well and form a thick, lush barrier or hedge. This shrub grows fairly fast, and adapts pretty well to growing in a large decorative container on a sheltered patio. Water regularly throughout the year, less in winter. Trim the plant to keep it neat, and nip out the growing points of young plants to encourage bushiness.

NATURAL DISTRIBUTION
High altitudes, on forest margins, exposed mountainsides and along streams.

> **HINT:** *Shrubs planted as hedges or screens slow the wind by diffusing it. Walls don't stop the wind, they merely deflect it (change its course of direction).*

2,5 m × 1 m

 to

Mainly spring, but throughout the year

Curry Bush ■ Small-leaved Curry Bush, Kerriebos, Kleinblaarkerriebos, Mokhwibitšana (N.So)

Hypericum revolutum CLUSIACEAE

Description and uses
Attractive bright-yellow flowers, each with a fluffy tuft of stamens in the centre, grace this fast-growing shrub with arching branches. The narrow pointed leaves of the Curry Bush are neatly arranged along the stems. They give off a distinct smell of curry after rain or when crushed.

Propagation
Propagate it from cuttings, or by removing young suckers from the root area.

Cultivation
This tough and beautiful shrub fares well anywhere in South Africa – it is very adaptable and easily grown in any soil. Of course, much better results are obtained when it is planted in good fertile well-drained soil with plenty of compost and sufficient water. Ideal for both small and large gardens. Experiment with a plant in a large decorative container on a sunny patio, where the pretty flowers and curry-scented leaves can be appreciated at close hand. Use the Curry Bush in an informal shrub border, preferably in full sun – but it will tolerate light shade for part of the day. Allow it to be the centre of attraction in a small townhouse garden. It tends to sprawl, so prune it back fairly hard at the end of winter to keep it neat and encourage new growth.

Natural distribution
High altitudes on forest margins (often a pioneer), in open grassland and along streams, from South Africa through tropical Africa to Ethiopia, Arabia and the Indian Ocean islands. NTN 484

Related species
Hypericum roeperianum (Golden Curry Bush, Large-leaved Curry Bush, Grootblaar-kerriebos) is a bushy rounded shrub (4 m x 3 m) with largish soft green leaves. It produces masses of deep golden-yellow flowers mainly in spring, with conspicuous tufts of stamens in the centre. Perfect for a pond or pool, where its flower-laden branches drooping softly to the ground are reflected in the water's surface. Adaptable and easy to grow, this *Hypericum* can be cultivated in much the same way as *H. revolutum*.

H. roeperianum

> **HINT**: Always feed and water plants in containers well, as they dry out quickly and run out of nutrients in their pots.

 Mainly spring, and any time during summer

3 m × 3 m

Wild Dagga ■ Lion's Ears, Wildedagga, Klipdagga, Duiwelstabak, Lebake (Sotho), imVovo (X), umFincafincane (X, Z), umCwili (Z)

Leonotis leonurus LAMIACEAE

Description and uses
Colourful and fast-growing, this perennial is drought-resistant and has long softly hairy tapering leaves with serrate edges. Bright orange nectar-rich velvety flowers are displayed in whorls at the tops of each stem. A form with creamy-white flowers is also available. Nectar-rich flowers entice butterflies, bees and birds such as the Whitebellied, Black, Yellowbellied, Olive, Collared and Marico Sunbirds. Traditional remedies prepared from the leaves heal colds, flu, coughs, bronchitis, headaches, asthma and high blood pressure. A leaf and root remedy is used against snakebite. A flower and leaf treatment deals with tapeworm. Twigs added to the bath are said to soothe itchy skin diseases and relieve muscle cramps. An infusion of the above-ground parts is applied to sores on the legs and head. Apparently it is also drunk as a slimming medicine! The earliest dwellers in South Africa chewed and smoked this plant like tobacco. Wild Dagga is not related to real dagga.

Propagation
Propagate it from seed, from cuttings or by dividing up large clumps. This is probably best done in early spring. Using a spade, lift the clump. Divide, and chop away the older less-vigorous looking sections. Replant newer, healthier-looking portions immediately into a well-prepared bed: don't leave clumps lying out in the sun for a long period of time. Water thoroughly immediately thereafter, and carefully for a couple of months until the plants are well established again.

Cultivation
It is very easy to grow and is not really fussy about soil, but will do best in a rich well-drained loam with plenty of compost added. Mulch well and replenish regularly. Water well in summer but keep almost dry in winter. Feed it with slow-release 3:1:5 fertiliser occasionally. Although this plant can get by with very little attention, it really flourishes when well cared for. Trim back whenever it looks a little untidy. Suitable for gardens of all types – use in an informal border, on a rockery, or closely planted in groups of 3–5 plants along a driveway. Because it is hardy, this low-maintenance plant is suitable for school gardens, university residences and any other similar situation, where upkeep must be kept to a minimum. Cut the plant right back at the end of winter – it will quickly send up new shoots.

Natural distribution
Amongst rocks in grassland.

> **HINT**: *The nectar-rich Wild Dagga is a perfect candidate for the 'bird garden'. Plants that flower in autumn, such as this one, help to tide nectar-loving birds over until the aloes begin to flower in early winter.*

2 m × 1,5 m

Autumn

Firewheel Pincushion ■ Vuurwiel-speldekussing

Leucospermum tottum PROTEACEAE

Description and uses
Leucospermums are attractive ornamentals that produce abundant and spectacular flowers. The Firewheel Pincushion is a neat compact rounded shrub with dark-green foliage and beautiful 'spiky' pink to brick-red 'pincushion' flowers. Each 'pincushion' is made up of many small flowers. Sunbirds such as the Malachite, Lesser Doublecollared and Orangebreasted Sunbird are attracted to the flowers and assist with pollination. The flowers are popular for flower arrangements.

Propagation
May be propagated from seed, but it is probably easier to buy plants from a nursery.

Cultivation
Protect young plants from frost for the first three years using an upturned box which must be removed in the morning, otherwise the plants may 'suffocate' and die (they need sun and wind movement). Moderately fast-growing, it requires acid well-drained soil (not clay!) with plenty of compost in a sunny, airy position, preferably on a bit of a slope. Natural koppies and hillsides are perfect because there is usually some wind and the run-off is good, so all you need is the sun! Provide a thick mulch using pine needles, straw, compost or pine bark, and do not apply artificial fertilisers or manure or disturb the root area by digging. *Leucospermums* are more suited to the winter rainfall area from which they originate, but will often do well in highveld gardens if correctly treated. In the south-western Cape, this attractive plant is drought resistant, but in the summer rainfall area it should be watered well both winter and summer. This is the perfect choice for an informal shrub border, or on slopes, banks or man-made rockeries where it will make a lovely show. The improved drainage on the slopes will definitely be to its liking. A really beautiful plant to have, if you can provide the right conditions. If you do not intend collecting the seed for propagation purposes, cut off the spent flowerheads after flowering. The bush can be trimmed back a little at the same time if necessary. *Leucospermums* have a lifespan of about 10 years.

Natural distribution
South-western Cape.

Oct–Jan

3 m × 3 m

Forest Bells ■ Mackaya, River Bells, Bosklokkiesbos, Blouklokkiesbos, uZwathi (Z), umAvuthwa (Z)

Mackaya bella　　　　　　　　　　　　　　　　　　　　　　　　　　　　　　　　ACANTHACEAE

DESCRIPTION AND USES
Forest Bells has lovely glossy dark-green leaves and is a wonderful shrub for shady areas. It is outstanding in full bloom, showered with exquisite pale mauve trumpet-shaped flowers. This plant originates in mountain forests and is adapted to low light conditions. Amazing patterns are visible in the throat of the flower under ultra-violet light – these are 'landing lights' which serve to guide the pollinator to the nectar supply in the flower under low light conditions. In this case the pollinator is a carpenter bee which makes its home in holes it bores in wood. The flower 'throat' or tube is exactly the right size for this large insect, a few species of which are found in our deeply shaded forests. The bush is heavily browsed by duiker. The wood is used to kindle fire by friction.

PROPAGATION
Easily propagated from cuttings.

CULTIVATION
This fast-growing shrub thrives when planted near water or in light shade under trees. Ideal for gardens both large and small. Use it as an ornamental in a large container on a shady patio – don't forget to water and feed it regularly, or it will not prosper. Perfect for the shady south side of a building, or use to decorate a shady entrance area. Suitable for use as a low but pretty screening plant. Mackaya prefers light fertile well-drained soil, such as good loam, with plenty of compost added. Mulch heavily – remember that this is a forest-loving creature that enjoys its home comforts – with lots of leafmould and humus! Water well in summer, but little in winter. Feed with slow-release 3:1:5 fertiliser at intervals of 6–8 weeks throughout summer. Pinch out the growing tips of young plants to encourage bushiness. Prune to shape and neaten when necessary. Use the prunings as mulch.

NATURAL DISTRIBUTION
Evergreen mountain forest, often along the edges of streams. NTN 681.1

3 m × 2 m

   to

Spring

Medium shrubs

Giant Honey Flower ▪ Touch-me-not, Kruidjie-roer-my-nie, Kriekiebos, Reuseheuningblom, Ubuhlungubemamba (X)

Melianthus major MELIANTHACEAE

Description and uses
A lush-looking ornamental with drooping blue-green foliage. Leaves are up to 75 cm long, attractively divided, with deeply toothed margins – when crushed, they give off a distinctive and unpleasant smell. Rusty-red flower spikes are carried high above the foliage and are followed by 'inflated' balloon-like fruits. Flowers are rich in nectar. Larvae of the Foxtrot Copper feed on this plant. Various leaf remedies using either *M. major* or *M. comosus*, are traditionally used to treat sores, bruises, septic wounds, ringworm, backache and rheumatism. Root infusions are traditionally used to treat cancer. *M. comosus* is used by the Khoikhoi to treat snakebite. Both *M. major* and *M. comosus* are poisonous and are known to have caused deaths through incorrect usage.

Propagation
Easily propagated from seed or rooted cuttings.

Cultivation
The Giant Honey Flower is a fast-growing vigorous invasive plant that may sucker and become a nuisance if not controlled – plant it where there is plenty of room for it to spread. Perhaps better suited to medium and large gardens. It makes a striking focal point planted alongside a stream or pond, or next to a modern-looking water feature, in an office complex garden. Suitable for a large informal shrub border. Try it in a big decorative pot on a patio. It is not really fussy about soil, but thrives if plenty of compost and organic material are added to the planting area. Feed with slow-release 3:2:1 fertiliser from time to time. Water fairly regularly in the growing season. Cut back to ground level if damaged by frost – it will rapidly shoot again in spring.

Natural distribution
Riverbanks and stream banks.

Related species
Melianthus comosus (Feathery Touch-me-not, Kruidjie-roer-my-nie, iBonya [Z]) is a fast-growing shrub (1,5 m x 1,5 m) with attractive dark-green feathery foliage and red nectar-rich flowers (sunbirds!), followed by inflated seed pods that become papery as they age. A decoction of the plant is applied to wounds that are healing too slowly. This plant grows in some pretty hot and dry places, so is suitable for gardens that have similar conditions. Use it in the shrub border and feed with slow-release 3:1:5 fertiliser from time to time.

M. comosus

Spring

2,5 m × 3,5 m

White Bristle Bush ■ Witsteekbos, Blombos, Lehlohlo (S.So)

Metalasia muricata ASTERACEAE

Description and uses
This well-branched rounded shrub sometimes forms a small tree. Narrow, needle-like grey-green leaves are either hairless or covered in woolly white hairs. Masses of white, pink, red or purple honey-scented flowers are displayed attractively in flattish heads at the tips of erect branches. Fruits are nutlets with bristles. The flowers attract a variety of insects, including the Painted Lady butterfly. They are also popular for flower arrangements. The White Bristle Bush is frequently used for stabilising coastal dunes. Stock such as sheep browse the foliage when food is scarce. In Lesotho a tea is prepared from the dried leaves. The Southern Sotho use this plant together with *Eriocephalus punctulatus* to fumigate a hut when someone has a cold, or after someone has died. Used as firewood in many areas.

Propagation
Propagate it from seed sown in March/April, or stem cuttings taken when the plant is actively growing. It sometimes seeds itself quite freely in gardens.

Cultivation
Hardy and fast-growing, the White Bristle Bush tolerates all climates and looks decorative for most of the year. It is not fussy about soil type and even grows well on the Cape Flats. Plant freely in coastal gardens – it is well adapted to these sometimes harsh conditions. In exposed areas it tends to form a more rounded bush, while in protected areas it can develop into a small bushy tree. This pioneer plant is perfect for newly established gardens where it can serve as a colourful temporary hedge or screen. Water moderately. Set plants out in groups of 3–5 in an informal shrub border, or position a single plant strategically in a small garden where it can be the centre of attraction when it flowers. Trim back after flowering to neaten.

Natural distribution
Coastal dunes, mountainous areas, rock outcrops and along streams. NTN 736

HINT: *Always add plenty of compost to the planting area when preparing flowerbeds, and mulch well after planting to protect the soil and reduce evaporation.*

3 m × 3 m

Any time, especially spring

Small-leaved Plane ■ Mickey Mouse Bush, Fynblaarrooihout, iLitiye (X), umBovu (Z)

Ochna serrulata OCHNACEAE

Description and uses
A decorative ornamental with beautiful pinkish-bronze spring foliage (later glossy green), large showy flowers, and the most unusual and intriguing fruits imaginable! Golden-yellow flowers, each with a tuft of fluffy stamens in the centre, are followed by shiny black fruits suspended below bright red sepals, looking just like Mickey Mouse faces! Fruit-eating birds are attracted by the fleshy fruits. Larvae of the One Spot Redwing moth feed on *Ochna* species. A root decoction is traditionally used by the Zulus to treat children with bone diseases.

Propagation
Propagation is difficult as the seed only germinates if very fresh, but cuttings of half-ripe shoots can be taken in summer. Wash seed and sow it (Dec–Jan) in well-drained loam seed compost (depth 1,5 × size of seed). Place in light shade and keep moist. Protect from rain. Germination time, 6 weeks; success rate, 90%. Young plants start flowering when they are about 50 cm high.

Cultivation
Plant out in good well-drained garden soil with plenty of compost and leafmould, apply mulch, and keep fairly moist throughout the year. Feed with slow-release 3:1:5 fertiliser in spring and early summer. The attractive Small-leaved Plane is suitable for an informal mixed border in gardens both large and small. If you are lucky enough to have natural rock outcrops, position a few of these plants strategically amongst the rocks. Use this shrub in a small townhouse garden where it can be the centre of attraction when it flowers and fruits. Try making a bonsai of it. This slow-growing plant happily copes at the coast, on the highveld, and in most other places in the country. Prune the *Ochna* lightly (especially when young) to give it a nice compact shape – it has a tendency to become a little untidy. Tolerates temperatures ranging from 5 °C to 35 °C and has a life expectancy of around 50 years.

Natural distribution
Forest margins and rocky hill slopes in South Africa and Swaziland. NTN 479.1

Sept–Oct

2,5 m × 2,5 m

Cape Leadwort ■ Plumbago, Blousyselbos, umaBophe (X, Z), umaSheleshele (Z)

Plumbago auriculata　　　　　　　　　　　　　　　　　　　　　　　　　　　　　　　PLUMBAGINACEAE

DESCRIPTION AND USES
A shrub with a tendency to scramble, the Cape Leadwort has small pale-green leaves and boasts masses of delicate powder-blue or white phlox-like flowers throughout summer. Butterflies are constant visitors in the warmer months, and the flowers attract the Greater Doublecollared Sunbird. Larvae of the Common and Short-toothed Blue butterfly feed on this plant. Poultry readily eat the leaves. The edible flowers are sometimes used to decorate cakes and salads. Traditional remedies prepared from the roots or leaves are used to relieve headaches, treat fractures and wounds, and remove warts. A decoction is drunk as a remedy for blackwater fever.

PROPAGATION
Propagate it from seed, cuttings or by lifting rooted suckers from the root area. Protect young plants from frost.

CULTIVATION
Use fertile, well-drained soil enriched with compost (lots and lots!) and water well in summer, but keep fairly dry in winter. Feed regularly with slow-release 3:1:5 fertiliser to keep it in peak condition. Allow it to scramble up amongst other shrubs or trees, or mass plant it on a sloping bank, possibly interplanted with yellow *Tecomaria* (see page 92) to provide a colourful combination. It makes an attractive informal hedge, or, if planted en masse, forms a good groundcover for a large garden. Fast-growing and fairly drought-resistant, this shrub may also be planted against a fence to form a screen or in an informal border where it can be kept neatly pruned. It does particularly well in frost-free areas, but if damaged by frost it will recover rapidly in spring.

NATURAL DISTRIBUTION
Scrub, thicket and valley bushveld.

3–5 m × 3 m

 to

Any time in summer

Purple Broom ■ Persbesem, Bloukappie, Ea moru (Nd), Hlokoa leleue (Sotho), Ithethe (Z)
Polygala virgata   POLYGALACEAE

DESCRIPTION AND USES
The free-flowering Purple Broom is a charming and beautiful shrub that would look marvellous planted anywhere – truly eye-catching. This fast-growing shrub has long narrow dark-green leaves and spikes of purple sweetpea-like flowers. Use the delicate flowers in a flower arrangement – they last well (for up to a week) in a vase. Leaves are heavily browsed. Bees pollinate the flowers. Traditional remedies prepared from the leaves and stems are used as blood purifiers.

PROPAGATION
Sow seed in autumn or spring – this plant seeds itself freely in the garden, so give a few of the small plants to a friend.

CULTIVATION
It tends to become a little 'leggy', so plant it behind other shrubs, e.g. *Felicia filifolia*, and always close together (about 60 cm apart), in groups of at least 3–5 plants. Young plants are very sparse, with single stems, but they fill out after a couple of years, to carry a dense, rounded crown of branches. The Purple Broom grows easily in any soil (preferably light and well-drained) that contains plenty of compost. It can be grown almost anywhere in South Africa, but in very hot areas position it so that it receives a little shade for part of the day. Water regularly and prune back severely if the plant becomes untidy – it will bush out again from that point. Regard these plants as temporary and replace every 3–4 years.

NATURAL DISTRIBUTION
Grassland and forest margins, from the Western Cape north to Tanzania and the Congo.

RELATED SPECIES
Polygala myrtifolia (Bloukappies, September-bossie) is a hardy and easily grown, rounded bushy shrub (3 m x 3 m) that produces pretty magenta-pink flowers all year. Carpenter bees are attracted to the flowers and the Laughing Dove eats the seeds. Use as an alternative to the Purple Broom, but remember to leave enough room for it to spread, as it is wider than that plant. In Gauteng, this plant is prone to attack by a mite that destroys the flowerbuds, thus spoiling the flowers. A granular insecticide that is sprinkled on the soil under the plant is supposed to help, but I did not find this to be the case.

P. myrtifolia

Jul–Sept

2 m × 1 m

King Protea ■ Giant Protea, Koningsprotea, Bergroos, Suikerkan

Protea cynaroides   PROTEACEAE

DESCRIPTION AND USES
This magnificent protea has the largest flowerhead of all the proteas, measuring nearly 30 cm across. It can vary in colour from pale cream-green to a soft, deep crimson-pink. An upright, rather compact shrub, the King Protea is usually only about 1 m tall, with broad leaves on red leafstalks. The flowers attract unbirds (e.g. Greater Doublecollared Sunbird) and sugarbirds (e.g. Cape Sugarbird). The Protea Canary eats the seeds. Larvae of the Orangebanded Protea butterfly and Protea Scarlet butterfly feed on this protea. The flower is an important export in the cutflower trade and also very popular locally for flower arrangements.

PROPAGATION
Propagation from seed is difficult – it is probably easier to buy plants from your local nursery. Plants flower when 4–5 years old.

CULTIVATION
Proteas are well worth growing, provided that you have suitable conditions. The King Protea will grow moderately fast in well-drained, acid soil – it will not succeed in heavy clay soil – in a sunny, airy position. Natural koppies and hillsides are perfect because there is usually some wind and the run-off is good, so all you need is a nice sunny spot. Water regularly throughout the year, especially in the summer rainfall area. Mulch well with pine needles, straw, compost or pine bark, but do not apply artificial fertiliser or manure and do not disturb the root area by digging. Stems terminating in spent flowerheads should be cut right back to ground level. This will encourage the plant to send up vigorous new flowering shoots. Adult plants are able to survive veld fires. This is one of the most adaptable and easy-to-grow proteas in cultivation and has been successfully grown in most moist parts of the country.

NATURAL DISTRIBUTION
Mountain ranges derived from Table Mountain Sandstone or Witteberg Quartzite. It grows right at the coast, where it is subject to salt-laden winds, and high up in the mountains (to 1 500 m), where it is sometimes engulfed in snow.

1,8 m × 1 m

May–Dec

Bearded Sugarbush ■ Woolly-bearded Protea, Baardsuikerbos

Protea magnifica PROTEACEAE

DESCRIPTION AND USES
The Woolly-bearded Protea forms a spreading rounded shrub, sometimes sprawling, with broad grey-green leaves. Its beautiful flowerheads, up to 20 cm across, are filled with a mass of soft white hairs, often tipped black or brown in the centre. Outer bracts vary in colour from soft pink to a deep rose. The Cape Sugarbird not only feeds on the sweet nectar in protea flowers, but also likes to nest in these plants. The Protea Canary eats the seeds.

PROPAGATION
Propagation from seed is difficult – it is easier to buy plants from your local nursery. Young plants do not usually flower before they are 6 years old.

CULTIVATION
This protea is slow-growing and requires extremely well-drained acid soil in a sunny airy position. It will not succeed in heavy clay soil unless plenty of river sand and compost are added, and the bed is raised, e.g. in a rockery or on a very steep slope. Natural koppies and hillsides are perfect because there is usually some wind and the run-off is good, so all you need is a nice sunny spot. Good drainage is an essential prerequisite to succeed with this particular protea. It thrives high on hot and dry mountains in summer, and survives freezing temperatures and snow in winter – all the while flowering unperturbed! In cultivation it tolerates light frost, but not humid, very moist summers. Water regularly in winter in the summer rainfall area. Mulch well with pine needles, straw, compost or pine bark, but do not apply artificial fertiliser or manure and do not disturb the root area by digging. If you do not plan to harvest the seed for propagation purposes, cut off the old flowerheads after flowering. Trim the bush back a little at the same time, if necessary.

NATURAL DISTRIBUTION
Mountains in the south-western Cape, where it is usually encountered at elevations of between 1 200 and 2 700 m. It favours dry, exposed sites that are hot in summer and cold in winter. NTN 86.1

> **HINT:** *A thick mulch layer helps to prevent weeds from germinating and will protect the soil surface from erosion. As it decomposes nutrients are returned to the soil for the plant's benefit.*

Jun–Jan

2 m × 2 m

Ruttyruspolia

X Ruttyruspolia 'Phyllis van Heerden'

ACANTHACEAE

DESCRIPTION AND USES
This attractive semi-deciduous shrub has softly textured leaves and a slightly sprawling to scrambling habit. Masses of striking pink-mauve to rose-pink flowers are freely borne in showy spikes in summer.

PROPAGATION
Propagate from cuttings.

CULTIVATION
Perfect for a large shrub border, where it can happily scramble into surrounding trees and shrubs. Moderately fast-growing: keep it in check by pruning or, if preferred, capitalise on its tendency to scramble, to show this beautiful shrub off to advantage. Use it in a large decorative container on a sunny patio – don't forget to water and feed it regularly. Groups of 3–5 plants, set out close together, can form a fairly effective but low screen. Free-flowering, it is useful for providing a bit of cheerful colour in medium to large home gardens, school grounds and shopping mall gardens. *Ruttyruspolia* is suitable for warmer gardens, and needs a light well-drained compost-enriched soil – the more compost the better. Apply a thick layer of mulch and replenish regularly. Feed with slow-release 3:1:5 fertiliser at intervals of 6–8 weeks throughout summer. Water well in summer, and prune the plant back after flowering, or if damaged by frost. This plant is a bi-generic hybrid between *Ruspolia hypocrateriformis* and *Ruttya ovata*.

RELATED SPECIES
Ruspolia hypocrateriformis var. australis (Red Ruspolia, Rooiruspolia) is a large rounded shrub (2,5 m × 2,5 m), with light-green softly textured leaves. Free-flowering, it boasts scarlet flowerspikes in summer. It tends to scramble into surrounding vegetation, but not as high as *Ruttyruspolia*. Moderately fast-growing, it is perfect for warmer gardens. In a big wildish garden allow it to clamber into nearby trees, or in a smaller suburban garden, keep it neatly pruned. Prune back hard after flowering, or if damaged by frost. *Ruttya ovata* is a rounded bushy shrub (1,8 m × 1,8 m), also inclined to scramble, that has tall spikes of white flowers. Cultivate both these shrubs as for *Ruttyruspolia*.

Ruspolia hypocrateriformis var. australis

3–4 m × 3 m

Summer

Cape Honeysuckle ■ Kaapse Kanferfoelie, Trompetters, Morapa-šitšane (N.So), Mothabathabane (N.So), Molaka (N.So), Malangula (Sw), umKoto (X), umUnyane (Z)

Tecomaria capensis (= *Tecoma capensis*)   BIGNONIACEAE

DESCRIPTION AND USES
Glossy dark-green foliage frames spikes of lovely yellow, orange, red or salmon trumpet-shaped flowers. Rich in nectar, they attract honeybees, butterflies (e.g. Pea Blue, Smokey Blue), sunbirds (e.g. Collared, Black, Marico and Scarletchested) and the Cape Sugarbird. The yellow-flowered form of the Cape Honeysuckle seems to form a neater, less sprawling bush. An excellent plant for the 'bird garden' – insectivorous birds will hunt the insects. Stock and game browse the leaves. Powdered bark is used for bleeding gums, and is said to relieve pain, fevers and influenza, and to induce sleep. A leaf decoction is used for diarrhoea.

PROPAGATION
Propagate it from cuttings or by lifting rooted runners or suckers from the root area, and planting them in pots. Do not transfer them into the garden until well established in the pots. Take softwood or herbaceous cuttings from a healthy vigorous mother plant in October. Treat with Seradix 2. Root in a bark and polystyrene mix (6:1), with bottom heat (28 °C) and mist. Rooting time, 12 days; hardening off, 2 weeks; success rate about 90%. During hardening off time, feed to encourage root development. Sow fresh seed (depth 1,5 x size of seed) from Nov to Mar. Use a well-drained loam seed compost. Place in 40% shade, protect from rain, keep moist. Germination time 2–3 weeks. Drench with fungicide after germination. Success rate 60–70%.

CULTIVATION
Plant the Cape Honeysuckle in any fertile well-drained garden soil (alkaline loam). Add plenty of compost, mulch well (replenish from time to time) and water regularly in summer. It has a tendency to spread and scramble, but may be pruned to keep it neat. This beautiful free-flowering shrub is fast-growing, drought-resistant and very easy to grow. With its non-invasive root system, it is ideal for gardens of all types. Plant it singly or in groups of three in a mixed border, on a golf course, or in large numbers to form an attractive informal hedge in warmer areas. If preferred and space permits, it can be allowed to scramble into surrounding trees or shrubs to show itself off to advantage. Wind-resistant, it grows well in coastal gardens. If temperatures drop too low it can lose its leaves. Feed with slow-release 3:1:5 in spring and again in early summer. Prune after flowering. Prefers temperatures between –3 °C and 35 °C and has a life expectancy of about 15 years.

NATURAL DISTRIBUTION
Margins of evergreen forest, in bush and scrub in coastal areas and along streams, from the Western Cape to Tanzania. NTN 673.1

Spring and autumn   2–3 m × variable

Large shrubs **93**

Weeping Sage ▪ Treursalie, Utile (X)
Buddleja auriculata LOGANIACEAE

DESCRIPTION AND USES
A lovely strong fragrance filling the air tells you that this shrub has anticipated spring long before you have – it comes as a real surprise in the middle of winter! The Weeping Sage has beautiful glossy foliage: its leaves are deep-green above and silver below. Profuse spikes of tiny tubular sweetly-scented cream, orange or lilac flowers appear on the ends of the 'weeping' branches. The fruit is a tiny creamy brown capsule that splits at the tip. The flowers attract numerous butterflies and other insects.

PROPAGATION
Easily propagated from hardwood cuttings.

CULTIVATION
Shapely and graceful, the Weeping Sage looks particularly attractive planted near water, perhaps next to a large dam or pond. The thick foliage provides safe shelter for birds. Use it in a large mixed shrub border, or to form a screen, or as an informal hedge. Suitable for medium to large gardens or any place that can accommodate its spread. This plant performs well in the Pretoria area, forming a neatish dense weeping shrub, but I am told that it does not fare as well in the Lowveld and possibly in other areas of the country – experiment to find out if this is true. Frost- and drought-resistant, this fast-growing shrub will probably grow well in most soils, but add plenty of compost and fertiliser (3:2:1 or 3:1:5) and water regularly for better results. Give less water in winter. It tolerates pruning well, but this is usually unnecessary if enough room has been allowed for it to spread comfortably.

NATURAL DISTRIBUTION
Mountain slopes, rocky ravines and forest margins, from the Eastern Cape to Zimbabwe. NTN 636.5

RELATED SPECIES
Buddleja salviifolia (Sagewood, Saliehout) has drooping silver-grey foliage (5 m x 4 m) and attractive terminal sprays of small white to cream or lilac flowers. Fragrant and rich in nectar, these flowers attract hordes of insects, including butterflies. Sagewood is semi-deciduous and makes a useful hedge or screen plant. It is not as dense as *B. auriculata* and *B. saligna*, and is slightly more untidy. Fast-growing, frost and drought-resistant, it is useful for new gardens. Prune well and often to keep it tidy. Cultivate as for the Weeping Sage. NTN 637

B. salviifolia

4 m × 4 m

Jul–Sept

Karoo Sage ■ Sneezebush, Karoosalie, Kakkerlak, Niesbos, Hoesbos

Buddleja glomerata LOGANIACEAE

DESCRIPTION AND USES
The Kakkerlak from the Karoo! This shrub displays its large spikes of tiny bright yellow flowers well above the attractive foliage. The flowers are strongly scented and apparently are said by some to smell like cockroaches, but they remind me of honey! Usually a bushy rounded shrub, this plant can also grow into a small tree 4 m high. The puckered wrinkled upper surface of the leaves is blue-green to silver-grey. The lower surface is covered by rusty to silvery white hairs. Leaf margins are deeply lobed. The leaves are often a brilliant silver-grey when growing in the Karoo mountains, where it makes an excellent foliage plant. In moister climates the leaves are not quite as silver. Young branches are pale green and so densely covered with silvery white scaly hairs that they appear rough or furry. The older brown wood appears to be covered in fine mauve powder that rubs off easily (actually a thin layer of scaly hairs). The fruit is a small hairy capsule.

PROPAGATION
Collect semi-ripe hardwood cuttings from an actively growing mother plant in October. Treat with Seradix 3. Root in sand, or a bark and polystyrene mixture (10:1), with bottom heat and mist. Rooting takes 30 days; hardening off, around 30 days; success rate about 45%. Not usually cultivated from seed.

CULTIVATION
A useful fast-growing shrub, resistant to heat and cold, that would make an attractive addition to any shrub border. Can be planted in gardens both small and large, wherever there is a suitable spot for it. Try using it in the drier parts of the country where its hardiness will be a distinct advantage. Use it as a hedge, windbreak or filler. Experiment with it in a coastal garden. Plant it on a natural koppie, or use to line a pathway – it has a non-invasive root system. Remember, allow room for it to spread, to prevent having to prune it often to remove offending branches. Grow it against the hot sunny west or north wall of the house. Plant it in a well-drained soil (acid or alkaline), such as a nutrient-rich loam, and add plenty of compost. Mulch well and water regularly until established. Thereafter, water moderately in summer, and do not overwater in winter. In moister areas, take care not to overwater it. Give it a good dose of compost each winter/early spring and feed with slow-release 3:2:1 or 3:1:5 fertiliser. Prune to neaten once in a while. Tolerates temperatures between –5 °C and 30 °C and has a life expectancy of about 17 years.

NATURAL DISTRIBUTION
Amongst rocks on hills and mountains in the Karoo. NTN 636.1

 Sept–Mar

4 m × 3 m

Large shrubs **95**

False Olive ■ Mock Olive, Witolien, Motlhwaretshogwana (Tsw), umGqeba (X), iGqeba-elimhlope (Z), iGqeba (Z)

Buddleja saligna　　　　　　　　　　　　　　　　　　　　　　　　　　　　　　　　　　　　　　　LOGANIACEAE

Description and uses
The False Olive is a dense bushy shrub with long narrow grey-green leaves and terminal sprays of minute white to creamy flowers. Fragrant and rich in nectar, they attract hosts of insects, such as bees, butterflies and beetles – this shrub is very popular with beekeepers. Insectivorous birds, such as the Barthroated Apalis, Cape and Natal Robins, Greyheaded and Orangebreasted Bush Shrikes, Kurrichane Thrush, barbets and bulbuls will be lured by the insects. A leaf decoction is traditionally used to deal with coughs and colds. The hard heavy dark-brown wood makes excellent fence posts, fuel and small items of furniture.

Propagation
Easily propagated from seed or cuttings.

Cultivation
The False Olive grows easily in any soil – add plenty of compost for better results – and will tolerate fairly heavy pruning. Mulch well, and feed with slow-release 3:2:1 or 3:1:5 fertiliser at intervals of 6–8 weeks throughout the growing season. This hardy evergreen makes an excellent hedge, screen, windbreak or waterside plant. In moister, warmer areas it may even form a small tree up to 8 m high. Use it in smallish gardens where there is not enough room for a bigger tree. Perfect for medium to large home gardens that have a spot for it. Because it is robust and tough it is an excellent low-maintenance plant for schools, farms, university residences and shopping mall gardens where upkeep must be minimised. Fast-growing and drought-resistant, it has a non-aggressive root system, so it is safe to plant it close to pools, buildings and driveways – not too close though, otherwise it will block access and drop leaves into the pool. It is usually best to keep plants at least 3 m away from any sort of construction. Train it into a nice compact shape, or it could become leggy. It can even be clipped into a neat hedge. It has a very wide distribution in South Africa and should grow well in most areas.

Natural distribution
Dry hillsides, mixed scrub and wooded valleys, on the margins of forests, along streams and in coastal bush, from the Western Cape to Zimbabwe. NTN 636

4 m × 3 m

 to

Oct–Apr

Climbing Num-num ▪ Ranknoemnoem, Amatungulu (Nd), Mothokolo (N.So), Ntamunga (Sha), Murambara (Shona), Tlaba-dilebanye (Tsw), Murungulu (V), iHlazane (Z)

Carissa edulis APOCYNACEAE

DESCRIPTION AND USES
Dense and thorny, the Climbing Num-num has glossy dark-green leaves and masses of fragrant white, pink-tinged, jasmine-like flowers. Spines are simple and unforked, and the delicious oval fruits ripen to purplish black – they are slightly larger than those of *C. bispinosa*, but smaller than those of *C. macrocarpa*. They are used to make appetising jams and jellies, a tasty pinkish wine or a vinegar. Fruit-eating birds such as Purplecrested and Grey Louries, African Green Pigeon, Crested and Blackcollared Barbets, starlings and parrots love them. The sweet-smelling flowers attract pollinating insects and insectivorous birds – an excellent plant for the 'bird garden'. Browsers are partial to the leaves; monkeys and baboons enjoy the fruits. Larvae of the Anomalous Emperor, Oleander Hawk, Pleasant Hornet, Duster and Arrow Sphinx moths feed on *Carissa* species. The root was once used as bitters when softened in rum or gin. In West Africa, a piece of root is placed in a gourd to give water a pleasant taste. Traditionally, the root is used to deal with coughs, chest complaints and gastric ulcers. An abortifacient and tonic are also prepared from it, and it is said to restore virility!

PROPAGATION
Sow in river sand, pressing seeds into soil until flush with surface. Cover with a thin layer of sand (equal to seed size). Place in a warm spot and keep moist. Germination time 7–14 days. Transplant (2-leaf stage) to nursery bags, in a mixture of soil and compost (2:1). Protect young plants from frost for a couple of years.

CULTIVATION
To form an attractive and impenetrable hedge, plant the Climbing Num-num close together (about 1 m apart). If planted under trees, this fast-growing ornamental may scramble up into the tree-tops. The root system is non-invasive, so this shrub can be planted reasonably close to walls, but remember it spreads sideways! Perfect for use as a screen, as it is quite dense – if planted for this purpose, space plants fairly close together. Can be used in a range of gardens, from small to huge. Ideally suited for large areas of open garden where three shrubs planted close together will draw the eye of any visitor. Plant in good fertile garden soil, enrich with plenty of compost, and water regularly. Apply a thick mulch layer and replenish when necessary. Feed with slow-release 3:1:5 fertiliser periodically in the growing season. Once established this plant may grow rather vigorously – prune and shape whenever necessary to keep it neat.

NATURAL DISTRIBUTION
Warm woodland and scrub, from Limpopo Province to southern Arabia. NTN 640.4

 Spring

3 m × 3 m

Amatungulu ■ Big Num-num, Grootnoemnoem, amaThungula (X), umThungulu (Z)

Carissa macrocarpa APOCYNACEAE

DESCRIPTION AND USES
An attractive ornamental with glossy dark-green foliage and starry white, sweetly scented flowers, larger than those of *C. edulis* and *C. bispinosa*. Although the Amatungulu usually forms a dense thorny shrub, it may also grow into a small tree up to 4 m high. It has strong, stiff spines, once or twice forked, and large (up to 5 cm long) bright-red oval fruits that are delicious and rich in vitamin C. They are one of our nicest indigenous fruits and are eaten whole, pips, skin and all! An exceptional jelly is made from them. Baboons, monkeys, fruit-eating birds and other animals enjoy the fruits. A pink dye is obtained from the fruit.

PROPAGATION
Propagate it from seed. See *C. edulis* for tips on how to do this.

CULTIVATION
The Amatungulu is popular both for its excellent nutritious fruits and for its ability to form a dense impenetrable hedge – it is even cultivated in California! Set the plants out close together – about 1 m apart – to form the hedge, and prune if necessary. Fast-growing and wind resistant, it is tolerant of coastal conditions. Suitable for any size garden, small to large. Perfect for an informal shrub border where it can be planted in groups of three. Use it to form a low screen, or to decorate the shady side of a wall. It grows fairly easily in fertile light well-drained soil enriched with tons of compost! Mulch well and replenish regularly. Feed with slow-release 3:1:5 fertiliser at intervals of 6–8 weeks throughout the growing season. Water regularly. Prune to neaten whenever necessary, but the plant does not usually get out of hand.

NATURAL DISTRIBUTION
Coastal bush, coastal forest and sand dunes, from the Eastern Cape to Mozambique. NTN 640.3

RELATED SPECIES
Carissa bispinosa (Forest Num-num, Bosnoemnoem, Serokolo [Tsw], Murungulu [V], umVusankunzi [Z]) is smaller (about 2 m × 2 m) than the Amatungulu, and has the same cultivation requirements. Also tolerates coastal winds. It has glossy dark-green leaves, tiny white, sweetly scented, jasmine-like flowers (summer), small red edible fruits (about 0,5 cm diameter), and twice-divided spines. The roots are traditionally used to treat toothache. NTN 640.5

C. bispinosa

3 m × 2 m

 to

Sept–Jan

Large shrubs

Cat's Whiskers ■ Tinderwood, Tontelhout, Stinkboom, Mohlokohloko (N.So), Moswaapêba (Tsw), iNunkisiqaqa (X), umQaqongo (Z)

Clerodendrum glabrum VERBENACEAE

Rotheca myricoides

DESCRIPTION AND USES
Cat's Whiskers has a leafy, often drooping crown of dark-green foliage, which may be evergreen. In warm moist areas it may grow into a small tree up to 8 m high. A profusion of small white pink-tinged flowers attracts hosts of insects and the tree hums with activity! Beautiful butterflies flit here and there, over the bush, and onto the fragrant flowers which buzz with bees and beetles. Clusters of round whitish pea-sized 'berries' that are smooth and shiny are loved by the White-eye, Blackeyed Bulbul and other birds. Larvae of the Purple-brown Hairstreak and Natal Barred Blue butterfly feed on the tree. The hard white wood is used to start fires, and for building huts and fish kraals. Crushed foliage smells strongly and is said to repel beetles. It is added to milk to rid calves of intestinal worms. A leaf decoction is placed on animal wounds to prevent infection by parasites (e.g. maggots). A root infusion is used as an antidote to snakebite. The leaf is used as a cough and fever remedy.

PROPAGATION
Propagate it from seed or from cuttings.

CULTIVATION
Plant in fertile garden soil enriched with compost, mulch well and water regularly. Feed with slow-release 3:1:5 fertiliser at intervals of 6–8 weeks throughout the growing season. It is fairly fast-growing; use it in a shrub border, or planted in a lawn, fairly close to the patio, where all the insect and bird activity can be appreciated! Probably more appropriate for gardens that are medium to large in size. Suitable for home and school gardens, parks and office complex gardens in slightly warmer areas where it will flourish. It grows naturally on coastal dunes and in coastal forest so it should thrive under similar conditions. Prune to shape whenever necessary.

NATURAL DISTRIBUTION
Open woodland, along rivers, in coastal forest and bush, and on coastal dunes, from the Eastern Cape northwards into tropical Africa. NTN 667

RELATED SPECIES
A useful alternative is **Rotheca myricoides** (= *C. myricoides*) (Blue Cat's Whiskers, Bloukatsnorbos, umTyatyambane [X], umBozwa [Z]), which varies from a scrambling shrub (3–7 m high) to a small tree. Flowers are an attractive electric-blue to purple. Foliage is soft and finely hairy, and also smells strongly when crushed, like that of *C. glabrum*. Perfect for a large informal border. Edible fruits are taken as a remedy against skin complaints. Cultivation requirements are similar to those of the Cat's Whiskers. NTN 667.1

Sept–Jan

5 m × 6 m

 to  to

Bluebush ■ Star Apple, Monkey Plum, Bloubos, Karoobloubos, Motloumana (N.So), Monkganku (S.So), umCafudane (Sw), Motlhaja (Tsw), Muthala (V), umBhongisa (X), umBulwa (Z)

Diospyros lycioides — EBENACEAE

Description and uses
A dense bushy rounded shrub with small silky leaves that are often blue-green, hence the common name 'Bluebush'. It may be deciduous or evergreen, depending on the climate, and sometimes forms a multi-stemmed tree (up to 7 m tall) – a very variable species. The grey bark is smooth. Tiny fragrant (especially at night) creamy-yellow bell-shaped flowers (male and female flowers on separate plants) are followed by masses of attractive deep-red marble-sized berries. The flowers attract hosts of insects and insectivorous birds. An excellent bee plant. The fleshy fruits are eaten by a variety of birds (e.g. barbets, louries and mousebirds), dassies, monkeys and also by humans – they have a pleasant sweetish taste and are used to make beer. A valuable shade, shelter and fodder plant for stock and game. Roasted ground seeds were once used as a coffee substitute. The wood is used to build huts and to make spoons. A yellowish-brown dye is obtained from the roots. The bark is used for tanning skins. Roots are chewed and the bristly ends used as toothbrushes (considered to be of a superior quality!)

Propagation
Easily propagated from seed (soak in hot water overnight). Use a mix of river sand and compost (equal parts), or a commercial seedling mix. Press seeds lightly into the soil, cover with a thin layer of sand and keep moist. Germination time 7–20 days. Transplant into small individual containers at the 2-leaf stage. Use a well-drained compost-rich soil. Prick out carefully and water well after transplanting. Young plants grow fast; place in the garden when they are about 1 m high.

Cultivation
Although this hardy and adaptable plant is found countrywide in a variety of soils, for best results in your own garden plant it in a light well-drained soil, add lots of compost, mulch, feed (slow-release 3:1:5)

and water regularly. Appropriate for a range of gardens, from small to large, provided that the spread of the plant can be accommodated. Plant freely on farms and game farms; also suitable for schools, office complexes and parks. Perfect for a large informal shrub border, or a rocky hillside or koppie. Useful as a low hedge or screen. Prune to shape and neaten when necessary. Remember, only the female trees make the beautiful fruits.

Natural distribution
Rocky hillsides (often quartzite outcrops) or alongside streams, from the Western Cape to Zambia. NTN 605

4 m × 4 m

Sept–Dec

Pink Wild Pear ■ Persblompeer, Pienkdrolpeer, Mokhwidi (N.So), iBunda (Z)

Dombeya burgessiae STERCULIACEAE

DESCRIPTION AND USES
Large soft heart-shaped leaves and drooping clusters of white to pale pink flowers with a red heart distinguish the Pink Wild Pear. It forms a large bushy shrub or a small tree up to 4 m high. Black Rhino browse the leaves and the bark, which also provides a fibre. If you are lucky, larvae of the Reticulate Bagnest moth will feed on your *Dombeya*. These larvae pack together in clumps on the stems after eating all the leaves, then link into long chains when it is time to look for places to pupate. The long snake-like processions, hence the name 'processionary caterpillars', heading all over the garden will amuse adults and children alike. And don't worry about your shrub – it will grow new leaves.

PROPAGATION
Easily propagated from cuttings taken in spring and summer or from seed.

CULTIVATION
Attractive and fast-growing, this ornamental flowers for a long time and is perfect for medium to large home gardens, school grounds, parks and office complexes in warmer areas. In smaller gardens it could possibly be pruned to form a tree, provided there is enough room to accommodate the spread. In cold gardens plant the Pink Wild Pear in a protected north-facing position in fertile well-drained soil. In cooler climates it may be semi-deciduous to deciduous. Add plenty of compost and other organic material; water regularly in summer, but less in winter. Mulch well and replenish the layer now and then. Feed periodically with slow-release 3:1:5 fertiliser. Experiment with one of these shrubs in a large decorative container on a patio, where the pretty flowers can be appreciated at close hand, or plant it in an informal shrub border. Make it the centre of attraction in a smallish garden. Don't plant it in very deep shade – it will perform better if positioned in lightly dappled shade or sun. The leaves are sometimes torn by strong winds, so if you live in a very windy area, position it in a slightly sheltered spot. Prune to neaten if necessary.

NATURAL DISTRIBUTION
Forest margins, along streams, open woodland and rocky hillsides, from KwaZulu-Natal to Kenya and Uganda. NTN 468.1

 Apr–Jul

4 m × 4 m

     to

Kei Apple ■ Kei-appel, Wildeappelkoos, amaQogolo (Nd), Mahlono (Pedi), Motlhono (N.So), Mukokolo (Shona), Motlhôno (Tsw), umQokolo (Z, X)

Dovyalis caffra　　　　　　　　　　　　　　　　　　　　　　　　　　　　　　　　FLACOURTIACEAE

DESCRIPTION AND USES
Kei Apples are drought-resistant, spiny shrubs. They have oval waxy light-green leaves and small cream-green flowers that are followed by large rounded apricot-coloured fruits (only on female trees). These are tasty (slightly acidic), rich in vitamin C, and can be eaten fresh or made into a delicious jam or jelly (add grapes when making a preserve). Birds love them, especially Purplecrested, Knysna and Grey Louries, Blackeyed Bulbul and mousebirds. Antelope, monkeys and baboons also enjoy them. The leaves are browsed by game and goats. A perfect plant for farms and game farms – a valuable source of fodder. The flowers lure insects of all sorts – Kei Apples are excellent candidates for the 'bird garden'. The larvae of the African Leopard butterfly feed on this shrub as well as the Sagewood. This shrub is especially cultivated for its fruits, but harvesting is a little difficult because of the thorns. Young fruits are made into pickles. The Pedi mix the fruit juice with porridge. It is popular in many parts of the world as a hedge plant.

PROPAGATION
Easily propagated from very fresh seed. Dry the fruits in the shade, remove the seeds and sow in river sand, pressing seeds down until level with the soil surface. Cover lightly with a layer of fine sand, and keep moist. Seedlings transplant easily (start with small containers). Protect young plants from frost for two years. Hardwood cuttings will root in river sand. First treat with Seradix.

CULTIVATION
The Kei Apple is evergreen under favourable conditions and moderately fast-growing. It prefers good soil (sand or loam), with plenty of compost and organic material added. Feed with slow-release 3:1:5 fertiliser at intervals of 6–8 weeks throughout the growing season. Perhaps a little thorny for the smaller garden, but perfect for medium to larger gardens that have the space for it. Position it where the thorns will not interfere with everyday traffic. It can be closely planted to form a security hedge. Prune heavily, if necessary, to keep it neat.

NATURAL DISTRIBUTION
Hot dry country, open bush and wooded grassland, rocky koppies and the edges of dune forest (where it can reach 8 m), from the Eastern Cape to Malawi. NTN 507

4 m × 3 m

  to

Nov–Jan

Pistol Bush ▪ Pistoolbos, isiPheka (X), iHlwehlwe (X), uHlwalana (Z), Lothabe (Z)

Duvernoia adhatodoides　　　　　　　　　　　　　　　　　　　　　　　　　　　　　　　　ACANTHACEAE

DESCRIPTION AND USES
The Pistol Bush forms an attractive, usually rounded shrub with large dark-green leaves and showy fragrant white flowers streaked red. Under favourable conditions (warmth and moisture) it may grow into a small tree up to 6 m high. Brown velvety club-shaped fruits burst open with a loud explosive sound, hence the name 'Pistol Bush'. Carpenter bees apparently pollinate the flowers.

PROPAGATION
Propagate it from seed. Under good conditions in the garden it seeds itself freely – give a plant to a friend.

CULTIVATION
A wonderful shrub for those shady moist parts of the garden, and beautiful in flower. It is easily grown, has pretty foliage, and is perfect for home gardens (small to large), school gardens, hotel grounds, parks and office gardens, in warmish areas. It grows well in Pretoria, but in exposed positions can sometimes be burnt by frost. When the frost is over, cut back the dead leaves. Position it under large trees, or use on the shady south side of the house, or to decorate a shady entrance area. Try planting it next to a stream or dam. Always ensure that it has some shade and sufficient moisture, or it will not do well and the leaves will turn yellow. This fast-growing plant prefers light well-drained fertile soil – add plenty of compost, mulch well, and water regularly for best results. Feed with slow-release 3:1:5 fertiliser at intervals of 6–8 weeks throughout the growing season. Prune if necessary to shape and neaten. Do this after the plant has flowered, unless you want to harvest the seed for propagation purposes.

NATURAL DISTRIBUTION
Coastal forest, forest margins and along streams. NTN 681

RELATED SPECIES
Duvernoia aconitiflora (Lemon Pistol Bush, Geelpistoolbos) is a fast-growing and easy-to-grow bushy rounded shrub (3 m x 3 m) with masses of two-lipped, pale lemon-yellow flowers. Sunbirds visit the flowers regularly. Cultivation requirements are similar to those of *D. adhatodoides*, except that it prefers lighter shade or slightly more sun. It is a little hardier and does not need as much moisture. NTN 681.2

D. aconitiflora

Summer　　　　　　　3,5 m × 3 m

Puzzle Bush ■ Hottentot's Lilac, Deurmekaarbos, Omusepa (Her), Umthele (Nd), Sekgalo (S.So), Morôbê (N.So, Tsw), umXele (Sw), umBotshani (X), umKlele (Z)

Ehretia rigida

BORAGINACEAE

DESCRIPTION AND USES
This many-stemmed shrub has a slightly 'weeping' habit – its arching branches spread and curve stiffly downwards, giving it a rigid, tangled and untidy appearance. Clusters of small sweetly scented lilac flowers are followed by edible, slightly sweet tasting berries which are orange-red to black when ripe. Yellowbilled Hornbill, Crested Barbet, Meyer's Parrot, Grey Lourie and other birds enjoy eating them as soon as they ripen. The delicate flowers lure butterflies, bees, beetles and other insects and, in turn, insectivorous birds. Leaves are readily browsed by stock and game. Use the Puzzle Bush as a source of fodder on farms and game farms. In Botswana bows are made from the pliable branches, which are used elsewhere to make fishing baskets. The wood is used for pestles. This tree is traditionally considered a good luck charm, and powdered root is used to treat gall-sickness in cattle. Another remedy deals with chest pains.

PROPAGATION
Easily propagated from seed or cuttings. For seed, use river sand, or a mix of river sand and compost (5:1). Sow seed (depth 1,5 x seed size). Press lightly into soil. Cover with a thin layer of sand (not too deep). Keep moist. Germination time, 10–20 days. Transplant into small bags at the 2-leaf stage. Place in the garden when about 1 m high.

CULTIVATION
The Puzzle Bush is excellent for a large informal border, or set plants out close together to form an informal hedge. Fast-growing, drought and frost resistant, it has been used in the centre of highways. Because it is so robust and hardy it is perfect for low-maintenance gardens of schools, shopping malls and university residences where upkeep must be minimised. However, always ensure that it has enough room to spread comfortably, otherwise time will be wasted pruning and shaping it to keep it within bounds! Appropriate for home gardens that are medium to large. Plant it in any compost-enriched soil and water moderately. Mulch well and replenish regularly. Fertilise from time to time in the growing season (slow-release 3:1:5) to keep it in good condition. It usually has to be pruned quite hard to keep it looking neat. It should do well in most parts of the country.

NATURAL DISTRIBUTION
A wide variety of habitats, such as evergreen forest margins (where it can reach 7 m), open woodland, on termite mounds and in coastal and karroid scrub, from the Western Cape to Zambia. NTN 657

4 m × 4 m

Spring

Eastern Cape Cycad ■ Oos-Kaapse Broodboom

Encephalartos altensteinii   ZAMIACEAE

Description and uses
Cycads are well-known ornamental plants with sturdy trunks, arching palm-like leaves and large cones. The Eastern Cape Cycad has bright-green leathery leaves up to 3,5 m long that are carried at the tops of often contorted trunks. Cones are golden yellow (male trees may have clusters of 3–5). In famine times, the starchy pith of the stem was used to make bread – hence the name 'bread tree' or 'broodboom'. Some cycad species are poisonous to man. Many fruit-eating birds (e.g. Trumpeter Hornbill), baboons, monkeys, bats and rodents eat the fleshy outer pulp of the seeds and later regurgitate or discard the poisonous hard-coated kernel. In the Eastern Cape the flesh of *E. altensteinii* seeds has been eaten by local people for generations.

Propagation
Purchase your plants from a cycad nursery.

Cultivation
E. altensteinii is moderately hardy and needs moderate watering. It tolerates a little shade, but grows better in sun. Cycads are tough and once established, do not need to be babied. Before planting, saturate the hole and allow the water to drain away. Never plant it any deeper or shallower than it is in its present container. You can add some compost. I feed my *E. lehmannii* and *E. villosus* every couple of months with 3:1:5. Don't overwater; apparently they like to have their leaves sprayed with water on hot days. Cut off old or dead brown leaves – but take note: they do not necessarily make new leaves every year.

Natural distribution
Coastal and inland areas of KwaZulu-Natal and the Eastern Cape. NTN 3

Related species
Encephalartos lehmannii (Karoo Cycad, Karoobroodboom (3 m × 3 m) – cycads with blue leaves, like this one from the Karoo and drier parts of the Cape Province, are the hardiest and need the least water. They are easy to grow, need full sun, and prefer neutral to alkaline soil. NTN 8.1. **Encephalartos villosus** (Ground Cycad, Grondbroodboom, Lisitsa [Sw], umGusa [X], Sehlati [Z]) – cycads with dark-green leaves, like this one, come from coastal and high-rainfall mountain forests. They need the most shade and water and are frost tender. This trunkless cycad (30 cm × 3 m) is easy to grow; leaves may reach 4 m in length.

> **NOTE:** Cycads are strictly protected and a permit is needed to own or move these plants.

7 m × 6 m

Large shrubs **105**

African Wild Banana ▪ Banana Palm, Afrika-wildepiesang, Motolô (N.So), Motholo (Pedi), Mulala (V)

Ensete ventricosum　　　　　　　　　　　　　　　　　　　　　　　　　　　　　　　　　　　　MUSACEAE

DESCRIPTION AND USES
The African Wild Banana has a stout fleshy stem crowned by lush tropical-looking banana-like leaves. Cream flowers are enclosed in maroon bracts, making a showy bunch up to 1,5 m long. The fruits look rather like bananas, but contain masses of hard black seeds resembling dried peas. It has a lifespan of about eight years – it flowers and fruits once, then dies. The young inflorescences (flowerheads) are edible and palatable when cooked. The fermented leaves are used to make bread. The dry insipid flesh of the fruits is eaten only in times of famine. The thick overlapping leafstalks are cut into small pieces and boiled, tasting like cooked celery. Stem and rootstock pulp residue is made into fibre and used for cordage and sacking.

PROPAGATION
Propagate from seed.

CULTIVATION
Plant the exceptionally fast-growing African Wild Banana in good light well-drained garden soil, add plenty of compost (and then some more!), and water well. Apply a thick layer of mulch to the soil surface, and replenish regularly. Feed with slow-release 3:2:1 fertiliser at intervals of 6–8 weeks throughout the growing season to help keep the plant looking lush and healthy. Don't forget to keep the soil moist. It is an excellent accent plant for a water garden and lends an immediate tropical atmosphere to wherever it is planted. Perfect for home gardens (large and small), office complexes, hotels, parks and school grounds in warmer areas. Cut off any untidy, drooping or dying leaves to keep it neat. The foliage will be destroyed in areas of severe frost, but the plant will recover rapidly in spring. Don't prune the plant until the frosts have ended – the burnt, dried leaves will help to protect the growing point.

NATURAL DISTRIBUTION
High-rainfall forest, mountain kloofs and near mountain streams. NTN 31

> **HINT**: *When using unrotted grass cuttings or clippings as mulch, feed the plants well with a nitrogen-rich fertiliser. This compensates for the nitrogen used when the grass decomposes.*

6 m × 4 m

   to

Honeybells ■ Honeybell Bush, Heuningklokkiesbos

Freylinia lanceolata SCROPHULARIACEAE

Description and uses
Golden-yellow honey-scented bells appear rather incongruous on this sometimes untidy shrub. Honeybells has long arching drooping branches of willow-like foliage. Usually multi-stemmed, it occasionally develops into a single-stemmed, weeping tree. The grey bark is smooth. Fruits are small brown capsules produced all year. Honeybells has a charm all of its own and attracts hosts of butterflies and other pollinators. Expect a visit from birds such as orioles, barbets and thrushes as well as Black and Whitebellied Sunbirds.

Propagation
Tiny wingless seeds germinate readily within three weeks. Take stem cuttings during the warmer summer months. Under suitable conditions young plants grow fast and may flower within a couple of seasons.

Cultivation
Add lots of compost to the planting area and mulch well. Water regularly, particularly if the shrub is planted in a herbaceous border away from water. It enjoys moist conditions and is very fast-growing if well watered (up to 80 cm per year). It would be perfectly at home positioned alongside a large dam, pond or water feature, where it could be kept pruned and tidied. If you have the time to spare, try pruning it into a single-stemmed tree. On farms, plant it on stream banks or in a large shrubbery, where the pretty flowers can be appreciated at close range. In home gardens, place it towards the back of an informal border – it is probably better suited to medium and larger gardens. Wind-resistant, frost-hardy and relatively pest-free, Honeybells fares equally well in summer- and winter-rainfall areas. Prune this adaptable plant whenever necessary to keep it neat. If you want to harvest seed for propagation purposes, don't cut off the old flowerheads.

Natural distribution
Moist areas along streams or on the edge of vleis. NTN 670.1

> *HINT: Draw a rough plan on paper before you start planting. Use circles to denote plant width. This will ensure that plants are well spaced and not overcrowded. Make the plan to scale using paper with blocks.*

Jun–Aug, and sporadically throughout the year

5 m × 5 m

Tonga Gardenia ■ Natal Gardenia, Natalkatjiepiering, Tongakatjiepiering, umValasangweni (Z), uNomphumela (Z)

Gardenia cornuta   RUBIACEAE

DESCRIPTION AND USES
The Tonga Gardenia is a dense rounded shrub with glossy light-green leaves, usually decorated with large shiny oval fruits that ripen to yellow. The lovely scent of the big showy white flowers fills the air in spring, and makes this a wonderful choice of ornamental for the home garden. Monkeys eat the young fruits. Leaves are browsed by game. The fruit is regarded by some as famine food. The branches are used for fencing and for fuel. This shrub is often planted at the gates of Zulu homes to keep away evil spirits (the Zulu name 'umValasangweni' means 'close the gate').

PROPAGATION
Propagate it from seed.

CULTIVATION
It grows moderately fast and may be planted in a mixed shrub border or alongside a pond or stream. Its neat attractive shape and appearance make it suitable for varied purposes in the garden. Appropriate for small, medium and large gardens. Experiment with a plant in a large container – don't forget to water and feed it regularly. This plant prefers slightly acid soil and lots of water. Mix in plenty of compost (and then more!) when preparing the soil for planting. Mulch well and replenish regularly. It does not usually require too much pruning – allow enough room for it to spread comfortably. Remember, if you cut off the old flowers the pretty fruits won't form! While this plant is growing and establishing itself (fairly slowly) fill the gaps with temporary fast-growing plants like *Euryops virgineus* and *Polygala virgata*.

NATURAL DISTRIBUTION
Grassland, thicket and in open woodland, in South Africa, Swaziland and Mozambique. NTN 690.1

RELATED SPECIES
The beautiful **Gardenia thunbergia** (Starry Gardenia, Witkatjiepiering, Tshiralala [V], umKhangazi [X], umKhwakhwane [Z]) is a slow-growing shrub or a small tree up to 5 m high (width 3 m). Breathtaking in full flower when masses of large starry white fragrant flowers cover the tree (summer). Distinctive egg-shaped grey fruits are speckled with whitish encrustations. They are very hard and woody and remain on the tree for a long time. Easily propagated from seed or truncheons. Can tolerate some shade and is slightly more frost tender than *G. cornuta*. Prefers light well-drained acid soil. Cultivate as for *G. cornuta*. NTN 692

4 m × 4 m

Spring

Crossberry ■ Kruisbessie, Assegaibos, Lesika (S.So), Msosobiana (Nguni, Nd), Motshwarabadikana (N.So), Mokukutu (Tsw), umNqabaza (X), iLalanyathi (Z)

Grewia occidentalis TILIACEAE

DESCRIPTION AND USES
This fast-growing shrub or small tree may be evergreen under favourable conditions. It produces starry pink flowers with a central mass of fluffy stamens, followed by small edible four-lobed fruits that are enjoyed by birds (e.g. Speckled Mousebird, Blackeyed and Cape Bulbuls, Crested and Blackcollared Barbets). Used as a source of fodder on farms and game farms – leaves are browsed by stock and game. Larvae of the Rufous-winged Elfin and Buff-tipped Skipper eat the leaves. The sweet fruits are widely eaten in Africa – they are often picked and stored for later use. Boiled in milk they taste like milkshake. In Botswana people drink the juice of the crushed fruits either fresh or fermented. The Xhosa make assegai handles and the San make bows from the wood. Bruised bark soaked in hot water is traditionally used to dress wounds. Rootbark deals with bladder problems. A shampoo prepared from crushed bark and regularly used prevents hair going grey (so it is said!)

PROPAGATION
Propagate it from seed – fresh seed germinates best. Young plants transplant readily and grow fast.

CULTIVATION
Plant the fairly drought-resistant Crossberry in fertile compost-enriched soil and water well in summer. Always add lots of compost to the planting area for best results. Mulch well, and feed with slow-release 3:1:5 fertiliser at intervals of 6–8 weeks throughout the growing season. Use it in an informal shrub border to attract birds. It is not too big and can quite comfortably be planted in small- to medium-sized gardens where its pretty flowers and glossy foliage will show up well. Situate it where visiting birds can easily be observed. This plant will occasionally 'lean into' surrounding vegetation and then scramble upwards. At other times it forms a nicely rounded shrub, or a small tree. The root system is non-aggressive so it can be positioned alongside driveways and other structures, but usually one shouldn't plant anything too close – this will often provide work for someone later on! Prune whenever necessary to keep it neat.

NATURAL DISTRIBUTION
Scrambles in evergreen forest, or is a shrub or small tree on forest margins, in open woodland, coastal bush and wooded hillsides, from the Western Cape to Ethiopia. NTN 463

 Nov–Feb

5 m × 3 m

 to to to

Large shrubs **109**

Tree Fuchsia
■ Notsung, Lebetsa (S.So), Mothêbêrêbê (N.So), umBinda (Sw), Murevhe (V), umBinza (X), uNondomela (Z)

Halleria lucida   SCROPHULARIACEAE

DESCRIPTION AND USES
No 'bird garden' is complete without the Tree Fuchsia. Brimful of nectar, its 'shy' brick-red or orange-red fuchsia-like flowers shelter amongst the soft leaves of this shrubby tree. Foliage is bright green and the pale grey bark is rough and grooved. Birds dart into the bush and disappear, only to reappear again minutes later. The tubular flowers grow directly on the branches, and, rich in nectar, attract insects such as bees and butterflies, and birds like the Protea Canary and Olive Sunbird. The rounded fleshy fruits are edible and ripen to black. White-eye, Cape and Kurrichane Thrushes, Natal and Cape Robins, Blackcollared and Crested Barbets, Rameron Pigeon, Purplecrested and Knysna Louries, Fiscal Flycatcher and mousebirds find them irresistible! Ripe fruits taste sweet; they are eaten raw or can be stored for quite a while. The hard wood is used for tools.

PROPAGATION
Easily propagated from cuttings or seed. Clean the flesh off first – it inhibits germination.

CULTIVATION
The extremely hardy and adaptable Tree Fuchsia is fast-growing in good soil (not heavy clay) and with sufficient water. It has a non-aggressive root system and makes a wonderful hedge or screen plant. It even tolerates some shade, and in the Cape, some wind. It can be planted in the shrubbery in home gardens, in parks, schools and office complex gardens, and on farms and game farms (cattle and game browse the leaves). It usually develops into a rounded bushy shrub 3–4 m high, but can also become a small tree, depending on prevailing conditions. A yellow-flowered form is also available.

NATURAL DISTRIBUTION
Occurs in a variety of habitats – along streams and in swamp forest, on mountain slopes, on rocky outcrops, in

deep forest (where it can reach 18 m) and on forest margins from the Western Cape to Ethiopia. NTN 670

RELATED SPECIES
Halleria elliptica (Wild Fuchsia, Kinderbossie) is a shrub (2,5 m x 2,5 m) with pretty bell-like orange flowers. It is more frost-tender than *H. lucida*, but otherwise cultivation requirements are similar. NTN 670.2

6 m × 5 m

  to

May–Dec

Parasol Flower ■ Chinese-hat Plant, Wildeparasolblom

Karomia speciosa forma *speciosa* (= *Holmskioldia tettensis*)   VERBENACEAE

Description and uses
The Parasol Flower is a well-branched shrub or shrubby tree up to 5 m high with small triangular leaves. In summer, the tiny bluish-purple flowers are topped with pinky-mauve calyxes that look like Chinese hats. These pink flower masses attract hordes of insects, particularly bees, and many birds amongst them the Southern Boubou.

Propagation
Seed or semi-hardwood cuttings.

Cultivation
An attractive fairly fast-growing tree/shrub for warmer gardens, where it can be planted in an informal shrub border or in a lawn. It should make an excellent bonsai, as it has tiny leaves and delicate flowers. Experiment with it in a large container on a patio where the flowers can be admired at close range. Use it in a smallish garden where space is limited and there is not enough room for a larger tree. It loses its leaves in winter, so it will allow a little sun into the garden. Choose a variety of smaller shrubs to create the shrubbery. It probably spreads a little too widely to be used comfortably in a townhouse garden. In cold gardens, plant this shrub in a protected position in well-drained sandy soil mixed with plenty of compost, and water well in summer. It prefers a warm moist summer and a dry mild winter. In the winter rainfall area it must be planted in a very well-drained position. Always apply a thick layer of mulch, and feed from time to time with slow-release 3:1:5 fertiliser in the growing season. Prune regularly to keep it neat.

Natural distribution
Riverine bush and open woodland, forest margins and rocky koppies, from KwaZulu-Natal northwards into tropical Africa. NTN 668

 Summer

4 m × 5 m

Rocket Pincushion ■ Perdekop

Leucospermum reflexum　　　　　　　　　　　　　　　　　　PROTEACEAE

Description and uses
Stunning salmon to orange rocket-shaped 'pincushion' flowers are displayed high above the attractive soft grey foliage of the Rocket Pincushion. Its nectar-rich flowers attract insects, sugarbirds and sunbirds. The flowers are popular for flower arrangements.

Propagation
Seed. Unless carefully handled seedlings damp off easily, so it may be better to purchase your plants from a nursery.

Cultivation
Protect young plants from frost; older plants are fairly tough. An excellent ornamental for the garden, and one of my favourite pincushions. Adaptable, it will grow in many areas of the country. Fast-growing and spreading, it prefers acid well-drained soil in a sunny airy position. Rocky koppies and hillsides are perfect as runoff is good and there is usually a little wind too. All you need is a sunny spot. My plant even thrived on a well-raised bank. Use it in an informal shrub border or on a large natural rockery where it can be seen to advantage. Suitable for steeply sloped roadside banks where it will beautify an unsightly highway. In the winter rainfall area it can happily be used in home gardens (small and large), office complex gardens, schools and parks. Its soft grey foliage and striking flowers will create quite an impact! Water moderately in summer and winter. In the summer rainfall area, apply more water in winter. Do not water the foliage; rather apply water at the base of the plant. Mulch well with pine needles, straw, compost or pine bark, but do not apply artificial fertiliser or manure and do not disturb the root area by digging. If you are not harvesting the seed for propagation purposes, cut off the old flowerheads after the plant has finished flowering. Trim back slightly at the same time if necessary.

Natural distribution
Along streams or in seepage areas in hot, arid mountain fynbos in the Clanwilliam area. NTN 85.2

3 m × 3 m

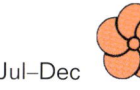

Jul–Dec

Large shrubs

Chinese Lanterns ▪ Klapperbos, Kiepkiepies

Nymania capensis   MELIACEAE

DESCRIPTION AND USES
This decorative slow-growing shrub has tiny leathery leaves. Pretty dark-pink bell-shaped flowers are followed by ornamental pink-red balloon-like seedpods that are papery and inflated. They would make an attractive and unusual addition to any flower arrangement – a wonderful opportunity to be innovative! The flowers attract sunbirds and honeybees.

PROPAGATION
Easily propagated from seed.

CULTIVATION
Perfect for hot dry gardens – it is drought resistant and can tolerate extreme heat and cold. In suitable areas, plant it in an informal shrub border, or use it to form a low hedge or screen. Position where its beautiful pods can be seen from the house or veranda. It can be sited on a large rockery, with a mix of attractive aloes, euphorbias and other succulents. Remember to harvest seeds if you wish to propagate the plant. Use it in home gardens (small to large), on farms and in rest camps at reserves. Chinese Lanterns prefers poor, well-drained soil and is difficult to cultivate on the highveld, although it may survive for a couple of years in a container placed in a dry position. It cannot tolerate overwatering or too much rain, so water sparingly. Try planting it on a very steep slope or rocky hillside to ensure adequate drainage; it would then perhaps have a better chance of survival. This is such an attractive and interesting plant that it is worthwhile experimenting with it – perhaps you can find a way to succeed with it! Prune to neaten if necessary.

NATURAL DISTRIBUTION
Hot, dry areas all over the Karoo and northwards to Namaqualand and Namibia. NTN 295

HINT: *For best results with plants, picture their natural habitat and climatic conditions, and then imitate these as best possible.*

 July onwards

4 m × 2 m

Snuff-box Tree ■ African Dogrose, Tonga, Snuifkalbassie, Tongwaan (Sha), Mothotse (N.So), umTongwane (Sw), umThumgwa (Tso), Mutudzwi (V), umShungu (Z)

Oncoba spinosa   FLACOURTIACEAE

Description and uses
Slow-growing and spiny, with glossy dark-green leaves, this attractive shrub may grow into a small evergreen tree up to 5 m high under favourable conditions. New foliage is coppery pink. The large fragrant white flowers have a central mass of fluffy yellow stamens. The large decorative fruits are about 6 cm in diameter, have hard shells and ripen to reddish-brown. These contain an edible, unpleasant tasting, mealy pulp and many small seeds that contain an edible oil. The hard-shelled fruits are used to make rattles for dancers, armbands for children and also snuff boxes – hence the common name. The root is traditionally used to deal with dysentery and bladder complaints. The light-brown wood polishes well but is too small to be of any value.

Propagation
Propagate it from seed sown in September, or from cuttings.

Cultivation
Plant the Snuff-box Tree in light, well-drained soil, add plenty of compost and other organic material, mulch well, and water throughout the year. Feed from time to time with slow-release 3:1:5 fertiliser in the growing season. Positioned close together these shrubs make an excellent thorny hedge or barrier. Plant in an informal shrub border or at the edge of a large pond or stream. Perfect for medium to larger gardens in warmer areas. The lovely fragrant flowers guarantee that it will be an asset anywhere. Use it as a tree in gardens that do not have space for bigger trees. Suitable for farms, schools and parks (plant where the spines will not be a nuisance). Prune to neaten if necessary.

Natural distribution
Riverine forest (where it can reach 10 m), and along streams in hot, open woodland, where it sometimes becomes thicket-forming; from Mpumalanga northwards into tropical Africa and Arabia. NTN 492

5 m × 4 m

 to

Spring and summer

Porkbush ▪ Elephant's Foot, Spekboom, iGqwanitsha (X), isiCococo (Z)

Portulacaria afra PORTULACACEAE

Description and uses
Most attractive in full bloom when it is a mass of soft pink, the Porkbush has a succulent glossy red-brown trunk and bears a dense crown of succulent leaves and stems. A profusion of small, starry pink nectar-rich flowers are displayed at the ends of branchlets, and are followed by tiny papery three-winged fruits. The nectar-rich flowers lure many insects (this is an excellent bee tree), in turn attracting insectivorous birds. The leaves are edible and have a pleasant acid flavour – in drier areas the plant provides excellent fodder for stock and game such as elephant and buffalo. In Mozambique breastfeeding mothers eat the leaves to increase their milk supply. In famine times, the Zulus eat the leaves raw. Larvae of the Diadem butterfly feed on *Portulacaria* species.

Propagation
Propagate it from cuttings, which must be kept fairly dry to prevent rot. Allow the cleanly cut edges of the cuttings to dry out for a few days before planting in well-drained river sand.

Cultivation
This shrub grows fairly fast and may be closely planted to form a hedge, used as an ornamental succulent tree for a rockery, or planted to check soil erosion. Use it as part of a plan to have only succulents in the garden. Suitable for dry areas where little else will grow. Easily grown, it is perfect for small to large-sized gardens, as well as parks, farms and reserves. It makes a beautiful bonsai. Remember when planting to use a well-drained soil that contains some bonemeal. The Porkbush can be planted in a large container on a sunny patio, but don't forget to water it when the container dries out. Feed the container with bonemeal (dig in with a small fork) and slow-release 3:1:5 fertiliser (about a tablespoon-full) at intervals of 6 weeks throughout summer. Prune when necessary to neaten.

Natural distribution
Dry, rocky koppies especially on Karoo hills and in succulent scrub. NTN 104

> **HINT:** *Well-chosen bushy spiny shrubs can form an impenetrable and unfriendly barrier, hedge or screen that will keep out most intruders.*

Oct–Nov

4 m × 3 m

Black Birdberry ■ Swartvoëlbessie, Mankgopo (N.So), Udzilidzili omhlophe (Sw), Tshidiri (V), umGonogono (X), iZele (Z), isiThitibala (Z)

Psychotria capensis RUBIACEAE

Description and uses
The attractive glossy foliage contrasts well with the rich creamy to golden-yellow flowers that are borne in flattish clusters at the tips of stems. Leaves are fairly large, smooth and leathery – dark green above and paler below. The Black Birdberry usually forms an evergreen shrub, but may occasionally develop into a small tree, often with a crooked trunk, and pale creamy-brown bark. Flowers are followed (in Jan–July) by bunches of rounded, fleshy, pea-sized fruits that ripen from yellow to red to blackish. Birds such as Blackeyed and Yellowbellied Bulbuls, Redwinged Starlings, robins and barbets enjoy them immensely. The hard, fine-grained yellow-brown wood is tough and makes a good general-purpose timber. It can be varnished and has a beautiful finish. Gastric complaints are traditionally treated with this plant.

Propagation
It grows fast and easily from seed. Wash the fresh seed. Sow (depth 1,5 x size of seed) in March and use bottom heat. Use a well-drained loam seed compost. Place in light shade and keep moist. Germination time, 6 weeks; success rate, 80%.

Cultivation
A native of the evergreen forest, it likes generous amounts of compost and other organic material – provide a thick mulch, renew regularly and water reasonably well in summer and under dry conditions. Soil must be an acid, well-drained, nutrient-rich loam. Feed with slow-release 3:1:5 fertiliser in spring and early summer. The Black Birdberry thrives when conditions are warm and moist and seems happiest in frost-free gardens. It is rather sensitive to the level of frost received in the Pretoria area – plant it in a very protected position in colder areas. Suitable for gardens of all sizes; it can be used where space is limited and there is not enough room for bigger trees. Roots are non-invasive, and it is perfect for shady areas under large trees and for an informal shrub border. Prune if necessary to neaten. The Black Birdberry prefers temperatures between 5 °C and 35 °C and has a life expectancy of about 50 years.

Natural distribution
Forest margins, evergreen forest (where it can reach 7 m), edges of rivers, scrub and dune bush and rocky outcrops in high-rainfall grassland, in South Africa, Zimbabwe and Mozambique. NTN 723

4 m × 3 m

     to

Aug–Jan

Dogwood ■ Blinkblaar, Umncaga (Nd), Mumbeza (Shona), Mokoyhi (So), Mofifi (N.So), Sineyi (Sw), umGlindi (X), umNyenye (X, Z)

Rhamnus prinoides   RHAMNACEAE

DESCRIPTION AND USES
Dogwood, with its beautiful high-gloss dark-green foliage, may form a dense, bushy shrub – sometimes scrambling into surrounding plants – or a small tree up to 6 m high. Small greenish flowers are followed by round pea-sized fleshy fruits, which ripen to purple. The flowers attract birds and insects such as bees – an excellent subject for the 'bird garden' and a good 'bee plant' for beekeepers. Starlings, bulbuls and barbets are some of the birds that may visit the garden to eat the fruits. Caterpillars of the Forest-king Charaxes and Pine Tree Emperor moth feed on the plant. The hard heavy white wood is suitable only for small articles, such as walking sticks. A root decoction is traditionally used to deal with pneumonia. A paste from the green leaves is applied to sprains. In Ethiopia, leaves are used as stimulants in wines and beers. It is believed that parts of the plant can protect against lightning.

PROPAGATION
Easily propagated from fresh seed – remove the fleshy part first, as it contains a growth inhibitor. Dry seeds in shade. Germination time, 2–6 weeks; success rate, 80%.

CULTIVATION
Dogwood grows easily in most soils and makes a lovely waterside plant or hedge (can be clipped). Fairly fast-growing, with a non-aggressive root system, it can safely be planted in smaller gardens and near paving or patios – not too close though, otherwise access may be blocked and you'll have a lot of pruning to do! This is another of those plants that loves its roots in plenty of leafmould; so add lots of compost and other organic materials to the planting area and mulch very well. Feed with slow-release 3:1:5 fertiliser at intervals of 6–8 weeks throughout the growing season. Plants growing in full sun seem to make sturdier and more rounded shrubs. Those in semi-shade tend to get a little lanky, are inclined to fall over and at the same time scramble up through other vegetation. Prune when necessary to neaten.

NATURAL DISTRIBUTION
Margins of evergreen forest, along riverbanks and in riverine bush, from the Western Cape to Ethiopia. NTN 452

> **HINT:** *The fleshy part of many fruits contains a germination inhibitor, so it is usually better to clean off the flesh before sowing or storing seed. Seed stored with flesh attached can become mouldy and lose viability.*

Oct–Dec

4 m × 4 m

Karoo Gold ■ Yellow Pomegranate, Geelberggranaat, Berggranaat, Ystervarkbos

Rhigozum obovatum   BIGNONIACEAE

DESCRIPTION AND USES
Karoo Gold is a twiggy spiny shrub with very small blue-green leaves and large golden-yellow trumpet-shaped flowers. It is spectacular in full bloom when it is a mass of bright yellow flowers. Fruits are flattened pod-like capsules. This is a valuable fodder plant (leaves, flowers and fruit) for stock and game in dry areas.

PROPAGATION
Easily propagated from seed. Collect the seed capsules when pale yellow brown, before they burst (Oct–Nov). Store in a dry well-ventilated room until they burst and release the seed. Store seed in dry airtight rustproof containers and dust with insecticide to protect it. Sow seed from Dec–Mar in a mixture of 3 parts gravel, 1 part rotted leaf mould (sieved), 1 part red loam. The medium must be very well drained, or seedlings will damp-off (rot). Scatter seeds far apart. Cover with a thin layer of sieved river sand or gravel. Water well, carefully. Place the tray out of harm's way, but in full sun to ensure that the seedlings are robust and strong from the start. Never allow the soil to dry out completely or become too soggy. Germination time 4–8 days. Transplant at the 4-leaf stage. Remove seedlings carefully, without damaging the fine hairlike roots. Transplant into small pots/black bags (roughly 500 ml) in a mix of 2 parts sieved gravel, 1 part red loam, 1 part sieved compost. Water well immediately and place in full sun. After 9 months transfer to 3 kg bags; thereafter, into the garden.

CULTIVATION
Karoo Gold is suitable for a range of gardens, from small to large. Young plants grow slowly until well established. Plant in a well-drained position, on a rockery, on a steep well-drained bank or slope, or in a mixed shrub border with other plants that require less water. Experiment with it in a large decorative container on a sunny patio, or next to a pool or driveway. Remember to water and feed the container plant regularly, but allow the soil to dry out fairly well between waterings. Add bonemeal to the well-drained potting mixture, and feed the plant with tiny sprinklings of slow-release 3:1:5 fertiliser at two-monthly intervals. Try making a bonsai from this plant. Karoo Gold grows moderately fast and has a non-aggressive root system. It can look a little dull for a good part of the year, but the beautiful flowers make up for it! Although this shrub tolerates drought and neglect, the results are better if plenty of compost is added to the soil, which should be gravelly. Heavy loam or clay soils seem to retard growth. In milder climates, the plant may retain its leaves. Prune to shape or neaten if necessary.

NATURAL DISTRIBUTION
Hot, dry, rocky areas in deciduous woodland and karroid vegetation, from the Western Cape to Namibia. NTN 675

3 m × 3 m

Sept–Nov

Dune Crowberry ■ Duinekraaibessie, Kraaibessie, umHlokotshane (X)

Rhus crenata ANACARDIACEAE

DESCRIPTION AND USES
A well-rounded bushy shrub with attractive glossy foliage. It may occasionally grow into a small tree up to 5 m tall. The bark is grey-brown and young branches are covered with soft reddish-grey hairs. The trifoliate (consisting of three leaflets) leathery leaves are dark-green, shiny above and paler below. Their margins are rolled under and the upper third is toothed. Young leaves are pinkish. The tiny creamy-white flowers are carried in short sprays at the tips of the branches. They are followed by clusters of slightly fleshy rounded berries that become red-brown and later blackish as they ripen (on the female plant). *Rhus* fruits are always popular with birds, so this is a useful shrub to plant if you want to encourage birds to your garden. Barbets, bulbuls, starlings and mousebirds are some of the types you can expect.

PROPAGATION
Propagate the Dune Crowberry from seed or cuttings. Remove the fleshy outer covering of the seed. Sow in a mixture of coarse sand, bark and loam (2:2:1). Can be placed in a cold frame. Success rate usually 50%. Plants are established more quickly from cuttings so this method is often preferred. Collect semi-ripe hardwood cuttings from a vigorous, actively growing mother plant in late winter or spring. Treat with Seradix 2 and Kaptan. Root in a mixture of bark and polystyrene chips (3:1), using bottom heat and intermittent mist. Rooting time about 8 weeks; hardening off, 2 weeks; success rate around 40%.

CULTIVATION
This fast-growing plant has non-invasive roots and develops into a neat spreading evergreen shrub. It is becoming increasingly popular for hedges, windbreaks and screening purposes. Plant it towards the back of an informal shrub border where it can form a nice backdrop. Perfect for a small garden that does not have room for larger trees and shrubs. Suitable for medium and large home gardens and office complexes where its neat appearance would be an advantage. It happily tolerates coastal conditions and hence is useful as a filler in coastal gardens and on the beachfront. When planting, always allow enough room for this shrub to spread. It would be a pity to spoil its shape by pruning it. Add plenty of compost to the soil (alkaline to neutral, well-drained sand) when planting and mulch well. Water regularly until the shrub is well established. The Dune Crowberry refers temperatures from 5 °C to about 25 °C and has a life expectancy of 15 years.

NATURAL DISTRIBUTION
On coastal (where it can reach 5 m) and inland dunes. NTN 380.1

 Feb–May

4 m × 4 m

Large shrubs **119**

Broom Karee ▪ Besembos, Besembessie, Ucane (Nd), Tselabele (S.So)

Rhus erosa ANACARDIACEAE

R. nebulosa

R. zeyheri

DESCRIPTION AND USES
The Broom Karee is a sprawling shrub or small tree up to 4 m high, with long narrow extremely finely toothed leaflets. It is an interesting garden subject, developing into a softly rounded shrub with a light and airy appearance. Just as some species of *Rhus* are cultivated for their bright and colourful fruits, others are cultivated for their ornamental foliage. *R. erosa*, *R. nebulosa* and *R. zeyheri* are good examples. Branches and leaves are used to make brooms, for fuel, cattle kraals and for thatching.

PROPAGATION
Collect the yellow-brown fruits in Feb/Mar or Apr/May. Place them in a bucket of water and discard those that float. Rub all the flesh off the fruits that sink to the bottom. Dry seeds in the sun. In Bloemfontein, it is best to sow in March. Use a mix of 1 part gravel, 1 part red loam, 1 part washed river sand. Don't sow too close together. Press seeds into soil lightly with a flat board. Cover with a thin layer (1,5 x seed size) of sieved gravel. Water well, place in full sun. Keep moist, never too dry and never too soggy. Germination time is 2 months. When seedlings have about four leaves, prick out carefully, keeping as much of the original soil around the roots as possible. Take care not to damage the hair roots. Pot into 500 ml containers in a mix of 1 part well-rotted compost (sieved), 1 part red loam, 1 part sieved gravel. Water well and place in a semi-shaded position. After 6 weeks, place in the sun. Keep relatively well watered at all times. When the plant is about 50 cm tall it can be planted into the garden. Suitable as a screening plant or as a focal point in small to large gardens. Water regularly until it is well established, whereafter it will be able to tolerate some drought. For a nice lush-looking plant, keep it well watered at all times. The Broom Karee plant can handle temperatures as low as –10 °C and as high as 42 °C.

NATURAL DISTRIBUTION
Rocky koppies. NTN 383

RELATED SPECIES
Rhus nebulosa* forma *nebulosa (Sand Currant, Sandtaaibos) is a semi-deciduous plant that forms a shrub, climber/scrambler or a small tree (2–4 m tall) with glossy dark-green foliage. The smooth and shiny attractive fruits ripen from yellow to bright red. An excellent pioneer plant, tolerant of coastal conditions and used for stabilising sand dunes. Used as a hedge at Kirstenbosch Gardens. It occurs naturally in dune scrub and on forest margins. NTN 390.1. ***Rhus zeyheri*** (Blue Currant, Bloutaaibos) is a beautifully neat and compact ornamental (3 m × 3 m) that is very under-utilised at present. It has attractive blue-green leathery foliage and occasionally develops into a small tree up to 4 m high. Its rounded decorative fruits ripen to russet-red. NTN 396.1

3 m × 4 m

Oct–Dec

Ulumbu Tree ■ Star-chestnut, Sterkastaiing, Koedoeklapper, uLumba (Sw), Samani (Tso), Mokgwakgwatha (N.So), Mokakata (Tsw), Mukakate (V), uLumbu (Z), iNkhuphenkhuphe (Z)

Sterculia rogersii　　　　　　　　　　　　　　　　　　　　　　　　　　　　STERCULIACEAE

Description and uses
Something really quite different for the garden – rather like a miniature baobab! The broad squat trunk branches from low down and the crown is rather sparse, making this a somewhat shrubby tree. The multi-coloured, succulent-looking trunk is very striking, and hence the Ulumbu Tree is an unusual addition to any garden. The grey bark is rough and tends to peel, revealing a beautifully smooth, dappled, creamy-grey to red or purple-mauve underbark. If you live in a frost-free area and like something extraordinary and beautiful – this is for you. The small yellowish bell-shaped flowers are streaked red. Fruits are star-shaped, with 3–5 lobes (or points). They start out greyish pink and velvety, later hardening and becoming golden brown. They then split, revealing dull blackish seeds resting in a bed of golden stinging hairs. The leaves are almost heart-shaped, shiny dark-green above, paler and softly hairy below. They usually fall early in autumn, baring the pretty trunk for all to see. Usually only about 3–4 m in height, this plant can reach 6–7 m under favourable conditions. Elephants browse the leaves and young stems. Kudus eat the leaves, flowers and fruits (before they ripen) – hence the name Koedoeklapper. People like eating the seeds, as do birds, game and stock. Twine or rope made from the bark is used in hut construction, to weave fishing nets and to sew sleeping mats. The rope is supposedly strong enough to drag away a dead kudu!

Propagation
The Ulumbu Tree is propagated from seed, truncheons and cuttings. Fresh seed germinates easily.

Cultivation
The plant prefers well-drained sandy soil, very hot summers and a dry temperate winter. Appropriate for medium to large gardens, it is equally at home on sandy flats and rocky hillsides. In moister, frost-free areas try planting it on slopes or banks to ensure good drainage. Perfect for large natural koppies or hillsides; suitable for rest camps in nature reserves, for example in Limpopo. A group of three in a large hotel or office complex garden would make a real impact! It may have an aggressive root system, so it is probably wiser to keep it away from buildings, pools and paving. Plant in very well-drained soil with plenty of compost. Water carefully (not too much!) until well established. Thereafter water less, and very little in winter.

Natural distribution
In dry bushveld, often on rocky koppies in South Africa, Swaziland, Botswana, Mozambique and Zimbabwe. NTN 477

Jul–Jan

5 m × 6 m

Natal Wild Banana ■ Natalse Wildepiesang, aNgude (X), Ikhamanga (X), iNkamanga (Z), Isigude (Z)

Strelitzia nicolai   STRELITZIACEAE

Description and uses
Palm-like, with tall bare stems crowned by large banana-like leaves, the Natal Wild Banana imparts an immediate tropical atmosphere to its surroundings. The large 'bird-of-paradise' flowers are purple-blue and cream, and produce abundant nectar that attracts sunbirds, in particular Olive, Whitebellied, and Black Sunbirds. Monkeys eat the soft parts of the flower and are partial to the orange, fluffy-looking seed aril. The larvae of the Banana Nightfighter butterfly feed on the leaves. The Zulus grind the seeds into flour, mix with water, and shape into small cakes, into which the oily orange arils are inserted. These cakes are then baked over coals and are said to be bland but filling.

Propagation
Propagate it from seed, or by removing suckers. In suitable climates these plants will seed themselves freely, so give the seedlings to a friend. Sow seeds (depth 1,5 × seed size) in a compost and river sand mix (equal parts). Press in gently and cover with a fine layer of sand. Keep moist. When robust, transplant the seedlings into small containers.

Cultivation
Wild Bananas are grown not so much for their flowers as for their special impact on the landscape, and their clean lines and shapes that suit them ideally to modern architecture. Often a single carefully placed plant can enhance the beauty of a garden. Perfect for medium to big gardens where they can be planted near large dams or water features. Position them where they will have the greatest impact. The root system is rather aggressive (keep away from buildings, foundations, etc.) and the plant tends to sucker from the base, so allow plenty of room for it to spread. Fast-growing, it prefers a protected position in frosty gardens, in fertile soil. Add plenty of compost and organic material to the planting area, mulch well, and water

regularly throughout the year in all areas. Feed with slow-release 3:2:1 fertiliser at intervals of 6–8 weeks throughout summer. Cut off any drooping, dying leaves to neaten. The large flowerstalk can also be removed after flowering if you do not want to harvest the seed for propagation purposes.

Natural distribution
Coastal bush and forest. NTN 34

8 m × 4 m

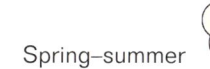

Spring–summer

Camphor Bush ■ Kanferbos, Vaalbos, umGebe (Sho), Setahlane (N.So), Mofahlana (S.So), Mohatlha (Tsw), isiDuli (X), iGqeba-elimhlophe (Z)

Tarchonanthus camphoratus ASTERACEAE

Description and uses
The Camphor Bush is either a dense bushy shrub or a small tree, often with a contorted trunk, occasionally up to 8 m high. The leathery grey-green leaves have an aromatic camphor-like smell. Terminal sprays of small creamy flowers are followed by tiny fruits covered in what looks like cottonwool – these white woolly sprays are very decorative. The root system is aggressive and this species can successfully be used to control soil erosion (binds sand on dunes). It is a valuable fodder tree, especially in arid areas – stock and game browse the foliage. The hard heavy grey-brown wood polishes well and is used for boat building, musical instruments, walking sticks and cabinet work – splinters are poisonous and cause septic sores. Poles make good termite-resistant fence posts, and firewood burns even when slightly moist. Zulu women use leaves to scent their hair. A leaf poultice is traditionally applied to deal with asthma and other chest complaints. Leaf infusions are used to treat toothache, headache, abdominal pain and bronchitis. Smoke of burnt branches relieves headaches and rheumatism. Leaves were apparently smoked like tobacco by the San and Khoikhoi people.

Propagation
Easily propagated from seed or softwood cuttings. Sow the seeds, wool and all! March and Sept–Oct are good times. Use a very well-drained mix that is not too rich: 1 part clean red soil, 3 parts gravel, 1 part coarse sand, 1 part well-rotted compost. The container must have adequate holes. Press the soil mix firmly into the container and level. Sow seed, pressing in well. Cover with sieved gravel. Germination unpredictable, 4–6 weeks. Keep seedlings moist, but don't overwater them. Transplant them at the 5–6 leaf stage. Water well the day before. Plant into 500 ml bags/pots using a mix of 3 parts clean red soil, 2 parts coarse sand and 1 part good fine compost. Take care not to damage the fine roots when transplanting. Do not overwater directly after transplanting, they may die off. Water carefully. After about six weeks, they should be ready to move to the next size container.

Cultivation
This shrub is perfect for gardens both small and large, provided that there is room for it to spread. It is wind resistant and very tolerant of coastal conditions. Fast-growing, drought-resistant and frost-hardy, plant it in any compost-enriched soil and water regularly. Prune to neaten whenever necessary.

Natural distribution
A wide variety of habitats, from arid semi-desert to moist coastal forest, from the Western Cape to Somalia. NTN 733

 Apr–Jun

5 m × 5 m
 to

Tulip Tree ■ Wild Tulip Tree, Wildetulpboom, umGeweba (Sw), iBicongo (Z), iPhuphume (Z)

Thespesia acutiloba — MALVACEAE

DESCRIPTION AND USES
Showy lemon-yellow flowers with semi-flared petals, striking red fruits and glossy triangular ivy-like leaves distinguish the Tulip Tree. This lovely spreading shrub or small tree, often branched from the base, has a grey trunk that becomes dark brown and deeply furrowed with age. Flowerbuds are red and open lemon-yellow at midday – they sometimes have maroon centres. Flowers can be up to 5 cm in diameter and often make their appearance when the tree is bare of leaves – stunning! The petals don't flare open fully, and the flowers close later in the afternoon. Eye-catching bright red, rounded (2 cm diameter) fruits sit in 'cups' (the calyx), that are carried on long stalks, reminding one a little of acorns. They are smooth and fleshy and ripen deep plum-red. Fruit-eating birds such as the Crowned Hornbill, Plumcoloured, Redwinged and Blackbellied Starlings, African Green Pigeon, Brownheaded Parrot and Purplecrested Lourie love them. This appealing plant is not well known, so is rather under-utilised at present. The tough blackish-brown wood is used for spear shafts, carving and musical instruments called 'makwelane'. A dye is obtained from the flower petals and the seeds are rich in oil. Roots are traditionally used in a bath to revitalise the body and drive off annoying spirits. The fibre has been used to make rope. The bark, root and fruit juice have been used as a remedy for cholera, dysentery, gall-bladder problems and for skin diseases.

PROPAGATION
Easily propagated from seed or cuttings.

CULTIVATION
The lovely Tulip Tree is well suited to waterside planting, and would enhance any large pond, dam or water feature. Ideal for the small to medium-sized garden and office complex garden, where its attractive foliage, flowers and fruits would be an asset. Perfect for larger gardens where it can be planted in a shrub border. Experiment with it in coastal gardens. Moderately fast-growing; add plenty of compost and other organic material to the planting area and mulch well. Water regularly until well established. As this tree makes an excellent garden subject, it is to be hoped that more people will now start to grow it. Well-established trees growing in the Pretoria National Botanical Garden have proved to be fairly drought tolerant during dry years. They also seem to cope with frost quite well. In bad frost years, leaves and young growth may be burnt, but the plant shoots again in spring. Prune to shape and neaten when necessary.

NATURAL DISTRIBUTION
Coastal and riverine forest (where it may reach 7 m), open woodland and fringes of mangrove swamps in South Africa, Mozambique and tropical Africa. NTN 465

4–6 m × 4–5 m

 to  to

 Feb–Apr

Honeysuckle Tree ■ Wildekamperfoelieboom, Kanferfoelieboom, umLahlana (X), umHlatholana (X), umLulama (Z), umAdlozane (Z)

Turraea floribunda　　　　　　　　　　　　　　　　　　　　　　　　　　　　　MELIACEAE

Description and uses
The Honeysuckle Tree is beautiful in early spring, when masses of attractive cream-green honeysuckle-like flowers appear on its bare branches. Their wonderful fragrance fills the air. Whitebellied and Black Sunbirds visit the flowers for their rich nectar, thereby assisting with pollination. The rounded deeply ribbed fruits split along the ribs into segments which curl back to expose their shiny bright orange-red seeds (the open fruits look like woody flowers). Fruit-eating birds such as Pied, Crested and Blackcollared Barbets, Grey and Purplecrested Louries, African Green Pigeon, Blackeyed Bulbul, Glossy Starlings and mousebirds are fond of the seeds and quickly remove them from the tree. Caterpillars of the Whitebarred Charaxes butterfly feed on the soft, light-green foliage. Traditional root remedies deal with dropsy, rheumatism, swollen joints and heart problems.

Propagation
Easily propagated from seed; young trees grow fast (up to 1 m in the first year).

Cultivation
An excellent ornamental for medium to large gardens, it may develop into a tree of up to 10 m high in warm, moist climates. In such climates, ensure before planting that there is enough room for it to spread. Perfect for an informal shrub border in a large garden. Place it where the pretty flowers and fruits – and all the visitors they attract – can be seen to advantage. Use it next to a large dam, or on the banks of a stream – it will look stunning. Plant it in a protected position in cold gardens, in fertile compost-enriched soil, mulch well, and water regularly in all areas. Feed with slow-release 3:1:5 fertiliser at intervals of 6–8 weeks throughout the growing season. Prune whenever necessary to neaten.

Natural distribution
Wooded ravines, open woodland, coastal bush (where it can reach 15 m) and along streams, from the Eastern Cape northwards into tropical Africa. NTN 296

 Sept–Nov

5 m × 3 m

 to

African Dogrose ■ Afrikaanse Hondsroos, umDubu (X), umBhalekani ■ isiShwashwa (Z)

Xylotheca kraussiana FLACOURTIACEAE

Description and uses
A beautiful shrub with glossy dark-green foliage. Attractive sweetly scented white flowers with a tuft of yellow stamens in the centre are followed by woody capsules with bright seeds. The African Dogrose has a long flowering season and grows moderately fast. Often a multi-stemmed shrub, it can form a small tree up to 7 m high in suitable climates. The fruit is a woody yellow capsule which narrows to a point and splits into 8 segments displaying sticky orange-red and black seeds, which birds and children love. Crested, Pied and Blackcollared Barbets, Redfaced and Speckled Mousebirds, Redwinged and Glossy Starlings, louries and bulbuls may be attracted to a fruiting tree. Larvae of the Rooibok and Blood-red Acraea butterflies feed on this shrub. Traditionally used to prepare love charms.

Propagation
Propagate from seed, and possibly from cuttings.

Cultivation
Plant the African Dogrose in a very protected spot in frosty gardens. It makes an excellent container plant for a sheltered patio, where the lovely flowers can be appreciated at close hand – remember to water and feed container-plants regularly. Ideal for gardens both small and large. The warmer and moister the climate, the bigger and more lushly the plant will grow, so take that into account when planning your garden. Suitable for an informal shrub border in a large warm garden. Perfect for hotel gardens and office complexes, where visitors can admire the pretty flowers. Experiment with it in coastal gardens – it grows naturally in coastal forest and on coastal dunes, so it should thrive. For the best results, plant this forest-loving shrub in light well-drained soil, combined with lots and lots of compost and leafmould; mulch well and water regularly. Feed with slow-release 3:1:5 fertiliser occasionally in the growing season. Prune whenever necessary to neaten.

Natural distribution
Coastal dunes, coastal forest (where it can reach 12 m) and forest fringes, sometimes on grassy slopes, in South Africa and Mozambique. NTN 493

3,5 m × 3 m

 to   to

Oct–Jan

References

Branch, B. 1990. *Bill Branch's field guide to the snakes and other reptiles of southern Africa*. Cape Town: Struik.
Burke, K.R. & Wolf, R. 1990. *Sunset, landscaping for privacy*. California: Lane Publishing.
Burrows, J.E. 1990. *Southern African ferns and fern allies*. Johannesburg: Frandsen Publishers
Cillié, B. 1997. *The mammal guide of southern Africa*. Pretoria: Briza.
Coates-Palgrave, K. 1988. *Trees of southern Africa*, 2nd ed. Cape Town: Struik.
Cowling, R. & Pierce, S. 1999. *Namaqualand – a succulent desert*. Vlaeberg: Fernwood Press.
Duncan, G.D. 2000. *Grow bulbs – Kirstenbosch Gardening Series*. Cape Town: National Botanical Institute.
Eliovson, S. 1973. *South African wild flowers for the garden*. Johannesburg: Macmillan.
Fabian, A. & Germishuizen, G. 1997. *Wild flowers of northern South Africa*. Vlaeberg: Fernwood Press.
Fox, F.W. & Norwood Young, M.E. 1988. *Food from the veld*. Johannesburg: Delta Books.
Giddy, C. 1974. *Cycads of South Africa*. Cape Town: Purnell.
Henning, G.A., Pringle, E.L.L. & Ball, J.B. (editors and revisers) 1994. *Pennington's butterflies of southern Africa*. Cape Town: Struik.
Herbarium Information — National Herbarium, Pretoria.
Hutchings, A.H., Scott, A.H., Lewis, G. & Cunningham, A. 1996. *Zulu medicinal plants, an inventory*. Natal: University of Natal Press.
Jeppe, B. 1969. *South African aloes*. Cape Town: Purnell.
Kirstenbosch Horticultural Notes.
Le Roux, A. & Schelpe, T. 1988. *Namaqualand, South African Wildflower Guide 1*. Cape Town: Botanical Society.
Maclean, G.L. 1996. *Roberts' birds of southern Africa*. Cape Town: John Voelcker Bird Book Fund.
Migdoll, I. 1987. *Field guide to the butterflies of southern Africa*. Cape Town: Struik.
Palmer, E. & Pitman, N. 1972. *Trees of southern Africa*. Cape Town: A.A. Balkema.
Palmer, E. 1985. *The South African herbal*. Cape Town: Tafelberg.
Pienaar, C. 1985. *Grow South African plants*. Cape Town: Struik.
Pinhey, E.C.G. 1975. *Moths of southern Africa*. Cape Town: Tafelberg.
Pooley, E. 1993. *The complete guide to trees of Natal, Zululand and Transkei*. Natal: Natal Flora Publications Trust.
Pooley, E. 1998. *A field guide to wild flowers of KwaZulu-Natal and the eastern region*. Natal: Natal Flora Publications Trust.
Rice & Rice, 1980. *Practical horticulture – a guide to growing indoor and outdoor plants*. Philadelphia: Saunders College.
Roberts, M. 2000. *Edible and medicinal flowers*. Cape Town: Spearhead Press.
Roodt, V. 1998. *Common wild flowers of the Okavango Delta – medicinal uses and nutritional value*. Gaborone: Shell Oil Botswana.
Roodt, V. 1998. *Trees and shrubs of the Okavango Delta – medicinal uses and nutritional value*. Gaborone: Shell Oil Botswana.
Rourke, J.P. 1980. *The proteas of southern Africa*. Cape Town: Purnell.
Schumann, D., Kirsten, G. & Oliver, E.G.H. 1992. *Ericas of South Africa*. Vlaeberg: Fernwood Press.
Sheat, W.G. 1982. *The A to Z of Gardening in South Africa*. Cape Town: Struik
Trendler, R. & Hes, L. 1995. *Attracting birds to your garden in southern Africa*. Cape Town: Struik.
Van der Spuy, U. 1971. *South African shrubs & trees for the garden*. Johannesburg: Hugh Keartland Publishers.
Van Wyk, B. & Gericke, N. 2000. *People's plants. A guide to the useful plants of southern Africa*. Pretoria: Briza.
Van Wyk, B., Van Oudtshoorn, B. & Gericke, N. 1997. *Medicinal plants of South Africa*. Pretoria: Briza.
Van Wyk, B. & Van Wyk, P. 1997. *Field guide to the trees of southern Africa*. Cape Town: Struik.
Van Wyk, B. 2000. *Photographic guide to wild flowers of South Africa*. Cape Town: Struik.
Walker, J. 1996. *Wild flowers of KwaZulu-Natal*. Pinetown: Walker family Trust.
Watt, J.M. & Breyer-Brandwijk, M.G. 1962. *The medicinal and poisonous plants of southern and eastern Africa*. London: E. & S. Livingstone.

Lecture notes: 'An Introduction to the Insects of Southern Africa', by Prof. Erik Holm.

Magazines:
Africa – Birds and Birding, volume 5, number 3, June/July 2000. Cape Town: Black Eagle Publishing.
Landbouweekblad, 5 March 1974.
Veld and Flora. Journal of the Botanical Society of South Africa. Cape Town: Creda Press. Many volumes used.

Index

ACANTHACEAE 36, 37, 38, 39, 49, 52, 57, 65, 70, 74, 83, 91, 102
Adenium multiflorum 32
African Dogrose 113, 125
African Wild Banana 105
Afrikaanse Hondsroos 125
Afrikaanse-salie 64
Afrika-wildepiesang 105
Agathosma betulina 33
Agathosma crenulata 33
Agathosma ovata **'Kluitjieskraal'** 33
Agtdaegeneesbos 51
Aloe arborescens 68
Aloe ferox 69
Aloe marlothii 69
Aloe tenuior 34
amaQogolo 101
amaThungula 97
Amatungulu 96, 97
ANACARDIACEAE 118, 119
aNgude 121
Anisodontea julii subsp. ***julii*** 35
Anisodontea scabrosa 35
APOCYNACEAE 32, 96, 97
Aromatic Sage 64
Asbossie 63
Ash Bush 63
ASPHODELACEAE 34, 68, 69
Assegaibos 108
ASTERACEAE 41, 45, 46, 47, 48, 63, 76, 85, 122

Baardsuikerbos 90
Banana Palm 105
Barleria albostellata 36
Barleria greenii 70
Barleria obtusa 37
Barleria pretoriensis 37
Barleria repens 38
Barleria repens **'Rosea'** 38
Barleria rotundifolia 39
Basoetokraal-aalwyn 34
Basterboegoe 33
Basuto Kraal Aloe 34
Bauhinia galpinii 71
Bauhinia natalensis 72
Bauhinia tomentosa 73
Beach Salvia 64
Bearded Sugarbush 90
Beesklouklimop 71
Beestebul 40
Belbos 67
Bergaalwyn 69
Berggranaat 117
Bergroos 89
Besembessie 119
Besembos 119
Bietou 76
Big Num-num 97
BIGNONIACEAE 92, 117
Bird-of-Paradise 66
Bitter Aloe 69
Bitteraalwyn 69
Black Birdberry 115
Blinkblaar 116
Blombos 85
Bloublom 64
Bloubos 57, 99
Blouheuningklokkiesbos 79
Bloukappie 88

Bloukappies 88
Bloukatsnorbos 98
Blouklokkiesbos 83
Blou-lippe 65
Blousalie 64
Blousyselbos 87
Bloutaaibos 119
Blue Cat's Whiskers 98
Blue Coleus 61
Blue Currant 119
Blue Honeybells 79
Blue Lips 65
Blue Salvia 64
Bluebush 99
Bobbejaanklou 50
Boetabessie 76
Bokbessie 53, 76
BORAGINACEAE 51, 103
Bosbeesklou 73
Bosboegoe 33
Bosklokkiesbos 83
Bosnoemnoem 97
Bosviooltjie 37, 38
Bridal Heath 44
Brillantaisia subulugurica 74
Broom Karee 119
Brown Salvia 64
Bruinsalie 64
Buchu 33
Buddleja salviifolia 93
Buddleja auriculata 93
Buddleja glomerata 94
Buddleja saligna 95
Buffelshoring 75
Burchellia bubalina 75
Bush Violet 37
Bushtick Berry 76

Cabhozi 60
Camphor Bush 122
Cancer Bush 67
Cape Gold 48
Cape Honeysuckle 92
Cape Leadwort 87
Cape Snowbush 45
Cargoe 53
Carissa bispinosa 97
Carissa edulis 96
Carissa macrocarpa 97
Cat's Whiskers 98
Chinese Lanterns 112
Chinese-hat Plant 110
Chrysanthemoides monilifera 76
Clanwilliam Euryops 46
Clerodendrum glabrum 98
Clerodendrum myricoides 98
Climbing Num-num 96
CLUSIACEAE 80
Crane Flower 66
Crassula ovata 40
Crassula portulacea 40
CRASSULACEAE 40
Crossberry 108
Curry Bush 80

Dainty Bauhinia 72
Deurmekaarbos 103
Didelta carnosa 41
Didelta spinosa 41
Dietes bicolor 42
Dietes grandiflora 43
Diospyros lycioides 99

Dogwood 116
Dombeya burgessiae 100
Douwurmbos 51
Dovyalis caffra 101
Dracaena aletriformis 77
Dracaena hookeriana 77
DRACAENACEAE 77
Dragon Tree 77
Drakeboom 77
Duinegousblom 41
Duinekraaibessie 118
Duiwelstabak 81
Dune Crowberry 118
Dune Marigold 41
Duvernoia aconitiflora 102
Duvernoia adhatodoides 102
Dwarf Coral Tree 78

Ea moru 88
Eastern Cape Cycad 104
EBENACEAE 99
Ehretia rigida 103
Eight Day Healing Bush 51
Elephant's Foot 114
Encephalartos villosus 104
Encephalartos altensteinii 104
Encephalartos lehmannii 104
Ensete ventricosum 105
Erica bauera 44
Erica glandulosa 44
Erica scabriuscula 44
Erica species 44
Erica versicolor 44
ERICACEAE 44
Ericas 44
Eriocephalus africanus 45
Erythrina humeana 78
Euryops speciosissimus 46
Euryops pectinatus 46
Euryops virgineus 47

FABACEAE 67, 71, 72, 73, 78
Fairy Iris 43
False Buchu 33
False Olive 95
Featherhead 58
Feathery Touch-me-not 84
Feë-iris 43
Firewheel Pincushion 82
FLACOURTIACEAE 101, 113, 125
Fluweelblaarbosviooltjie 36
Forest Bells 83
Forest Num-num 97
Forest Spurflower 60
Freylinia lanceolata 106
Freylinia tropica 79
Fynbeesklou 72
Fynblaarrooihout 86

Gansies 67
Gardenia cornuta 107
Gardenia thunbergia 107
Geelbarleria 39
Geelbeesklou 73
Geelberggranaat 117
Geelblomsalie 64
Geelpiesang 66
Geelpistoolbos 102
Geelsewejaartjie 48
GERANIACEAE 55, 56
Giant Honey Flower 84
Giant Protea 89
Giant Salvia 74
Golden Curry Bush 80

Golden Daisy 46
Golden Salvia 64
Gouemargrietjie 46
Green's Barleria 70
Grewia occidentalis 108
Grey Barleria 36
Grey Bush Violet 36
Grondbroodboom 104
Groot Wilde-iris 43
Grootblaardrakeboom 77
Grootblaarkerriebos 80
Grootnoemnoem 97
Ground Cycad 104
Grysbarleria 36

Halleria elliptica 109
Halleria lucida 109
Harigemalva 35
Harpuisbos 46
Heart-leaved Pelargonium 55
Heaths 44
Heide 44
Helichrysum patulum 48
Helichrysum splendidum 48
Heuningklokkiesbos 106
Heuningmargriet 47
Hlokoa leleue 88
Hoesbos 94
Holmskioldia tettensis 110
Honey Euryops 47
Honey Everlasting 48
Honeybell Bush 106
Honeybells 106
Honeysuckle Tree 124
Hottentot's Lilac 103
Hypericum revolutum 80
Hypericum roeperianum 80
Hypoestes aristata 49

iBicongo 123
iBonya 84
iBunda 100
iDololenkonyane 37
iGonsi-lasehlathini 77
iGqeba 95
iGqeba-elimhlope 95
iGqeba-elimhlophe 122
iGqwanitsha 114
iHlazane 96
iHlwehlwe 102
iKati 78
iKhala 69
iKhalene 34
Ikhamanga 121
iLalanyathi 108
iLitiye 86
Impala Lily 32
Impala-lelie 32
Impepho 48
imVovo 81
inHlaba empofu 34
inKalane encane 68
iNkamanga 121
iNkhuphenkhuphe 120
inTelezi 34
iNunkisiqaqa 98
Inyanga Hedge Plant 79
iPhuphume 123
IRIDACEAE 42, 43
isiCococo 114
isiDuli 122
isiGolwane 75
Isigude 121

isiGude 66
isiPheka 102
isiQungasehlati 42, 43
isiShwashwa 125
isiThibathibana 73
isiThibothi 65
isiThitibala 115
Ithethe 88
iThobankomo 75
iTholonja 76
iTokothoko 77
iZele 115

Jantjie-Bêrend 67

Kaapse Kanferfoelie 92
Kakkerlak 94
Kalkoentjiebos 67
Kanferbos 122
Kanferfoelieboom 124
Kankerbossie 67
Kapokbos 45
Kapokbossie 45
Karkey 40
Karomia speciosa forma ***speciosa*** 110
Karoo Cycad 104
Karoo Gold 117
Karoo Sage 94
Karoobloubos 99
Karoobroodboom 104
Karoosalie 94
Kei Apple 101
Kei-appel 101
Kerriebos 80
Kiepkiepies 112
Kinderbossie 109
King Protea 89
Kisololo 71
Klapperbos 112
Kleinblaarkerriebos 80
Kleinbosviooltjie 38
Kleinkoraalboom 78
Klipdagga 81
Knoffelsalie 61
Koedoeklapper 120
Koningsprotea 89
Kooigoed 48
Kraaibessie 118
Kraanvoëlblom 66
Kransaalwyn 68
Krantz Aloe 68
Kriekiebos 84
Kruidjie-roer-my-nie 84
Kruisbessie 108

LAMIACEAE 54, 59, 60, 61, 62, 64, 81
Large Spurflower Bush 59
Large Wild Iris 43
Large-leaved Curry Bush 80
Large-leaved Dragon Tree 77
Laventelbossie 63
Lebake 81
Lebetsa 109
Lehlohlo 85
Lemon Pistol Bush 102
Leonotis leonurus 81
Lesika 108
Leucospermum cordifolium 50
Leucospermum reflexum 111
Leucospermum tottum 82
Lintbos 49

Lion's Ears 81
Lisitsa 104
Lobos 51
Lobostemon fruticosus 51
Lobster Flower 61
LOGANIACEAE 93, 94, 95
Lothabe 102
Lowveld Bauhinia 71
Luibossie 51
Luisiesboom 50

Mackaya 83
Mackaya bella 83
Mahlono 101
Mahlozana 75
Malangula 92
MALVACEAE 35, 123
Mankgopo 115
Mazabuka 65
MELIACEAE 112, 124
MELIANTHACEAE 84
Melianthus comosus 84
Melianthus major 84
Metalasia muricata 85
Metarungia longistrobus 52
Mickey Mouse Bush 86
Mmaba 53
Mock Olive 95
Mofahlana 122
Mofifi 116
Mogopa 69
Mohatlha 122
Mohlokohloko 98
Mokakata 120
Mokgwakgwatha 120
Mokhupye 78
Mokhwibitšana 80
Mokhwidi 100
Mokoyhi 116
Mokukutu 108
Molaka 92
Monkey Plum 99
Monkga-nku 99
Moraithama 57
Morapa-šitšane 92
Morôbê 103
Moswaapêba 98
Mothabathabane 92
Mothêbêrêbê 109
Mothokolo 96
Motholo 105
Mothotse 113
Motlempa 76
Motlhaja 99
Motlhono 101
Motlhôno 101
Motlhwaretshogwana 95
Motloumana 99
Motolô 105
Motshiwiriri 71
Motshwarabadikana 108
Mountain Aloe 69
Msosobiana 108
Muishondblaar 60
Mukakate 120
Mukokolo 101
Mulala 105
Mumbeza 116
Murambara 96
Murevhe 109
Murungulu 96
Murungulu 97
MUSACEAE 105
Muthala 99
Mutudzwi 113

Natal Bauhinia 72
Natal Gardenia 107
Natal Wild Banana 121
Natalkatjiepiering 107
Natalse Wildepiesang 121
Natalsebeesklou 72
Niesbos 94
Nodding Pincushion 50
Notsung 109
Ntamunga 96
Nylandtia spinosa 53
Nymania capensis 112

Ochna serrulata 86
OCHNACEAE 86
Omusepa 103
Oncoba spinosa 113
Oos-Kaapse Broodboom 104
Orthosiphon labiatus 54
Oval-leaf Buchu 33

Parasol Flower 110
Peacock Flower 42
Pelargonium cordifolium 55
Pelargonium graveolens 56
Perdeblom 41
Perdebos 41
Perdekop 111
Persbesem 88
Persblompeer 100
Persmuishondblaar 59
Petalidium 57
Petalidium oblongifolium 57
Phefo-ea-loti 48
Photsoloma 77
Phylica plumosa 58
Phylica pubescens 58
Pienkdrolpeer 100
Pienk-kiesieblaar 35
Pienksalie 54
Pink Fly Bush 60
Pink Joy 40
Pink Mallow 35
Pink Sage 54
Pink Wild Pear 100
Pistol Bush 102
Pistoolbos 102
Plakkie 40
Plectranthus zuluensis 62
Plectranthus ecklonii 59
Plectranthus fruticosus 60
Plectranthus neochilus 61
Plectranthus saccatus 62
PLUMBAGINACEAE 87
Plumbago 87
Plumbago auriculata 87
Polygala myrtifolia 88
Polygala virgata 88
POLYGALACEAE 53, 88
Porkbush 114
PORTULACACEAE 114
Portulacaria afra 114
Poublom 42
Pride-of-De Kaap 71
Protea cynaroides 89
Protea magnifica 90
PROTEACEAE 50, 82, 89, 90, 111
Psychotria capensis 115
Pteronia incana 63
Purple Broom 88
Puzzle Bush 103

Ranknoemnoem 96
Red Ruspolia 91
Reuseheuningblom 84
Reusesalie 74

RHAMNACEAE 58, 116
Rhamnus prinoides 116
Rhigozum obovatum 117
Rhus crenata 118
Rhus erosa 119
Rhus nebulosa forma **nebulosa** 119
Rhus zeyheri 119
Ribbon Bush 49
River Bells 83
River Resin Bush 47
Rivierharpuisbos 47
Rocket Pincushion 111
Rooiruspolia 91
Rose-scented Pelargonium 56
Rotheca myricoides 98
Round-leaf Buchu 33
RUBIACEAE 75, 107, 115
Ruspolia hypocrateriformis var. **australis** 91
RUTACEAE 33
Ruttyruspolia 91

Sabie Star 32
Sagewood 93
Saliehout 93
Salvia africana-lutea 64
Salvia chamelaeagnea 64
Salvia muirii 64
Salvia thermara 64
Samani 120
Sand Currant 119
Sandsalie 64
Sandtaaibos 119
Sclerochiton harveyanus 65
SCROPHULARIACEAE 79, 106, 109
Seegousblom 41
Seeroogblommetjie 49
Sefahlane 122
Sehlati 104
Sekgalo 103
Sekgopha 68
Septemberbossie 88
Serokolo 97
Shell Bush 54
Sineyi 116
Skilpadbessie 53
Slaaibossie 41
Small Bush Violet 38
Small-leaved Curry Bush 80
Small-leaved Plane 86
Sneezebush 94
Snuff-box Tree 113
Snuifkalbassie 113
Sonbekkiebos 52
Spekboom 114
Speldekussing 50
Spiny Yellow Barleria 39
Spoorsalie 60
Star Apple 99
Star-chestnut 120
Starry Gardenia 107
Sterculia rogersii 120
STERCULIACEAE 100, 120
Sterkastaiing 120
Stinkboom 98
Stoep Jacaranda 62
Stoepjakaranda 62
Strandsalie 64
Strelitzia nicolai 121
Strelitzia reginae 66
STRELITZIACEAE 66, 121
Suikerkan 89
Sunbird Bush 52
Sutherlandia frutescens 67

Swartvoëlbessie 115

Tapaalwyn 69
Tarchonanthus camphoratus 122
t'Arda 41
Tecoma capensis 92
Tecomaria capensis 92
Thespesia acutiloba 123
Thlakeni 40
Thorny Bush Violet 39
TILIACEAE 108
Tinderwood 98
T'kaibebos 63
Tlaba-dilebanye 96
Tonga 113
Tonga Gardenia 107
Tongakatjiepiering 107
Tongwaan 113
Tontelhout 98
Tortoise Berry 53
Touch-me-not 84
Tree Fuchsia 109
Treursalie 93
Trompetters 92
Tropical Salvia 74
Tselabele 119
Tshidiri 115
Tshiralala 107
Tswiriri 71
Tulip Tree 123
Turkey Flower 67
Turraea floribunda 124

Ubuhlungubemamba 84
Ucane 119
uDzilidzili omhlophe 115
uHladlwana olukhulu 49
uHlonyane 49
uHlwalana 102
Uiltjie 42
uLumba 120
uLumbu 120
Ulumbu Tree 120
uLwamfithi 76
umaBophe 87
umAdlozane 124
umaSheleshele 87
umAvuthwa 83
umBhalekani 125
umBhongisa 99
umBinda 109
umBinza 109
umBotshani 103
umBovu 86
umBozwa 98
umBulwa 99
umCafudane 99
umCwili 81
umDubu 125
umfincafincane 81
umFincane 75
umGebe 122
umGeweba 123
umGlindi 116
umGonogono 115
umGqeba 95
umGusa 104
umHlaba 69
umHlabana 68
umHlatholana 124
umHlokotshane 118
umKhangazi 107
umKhoma-khoma 77
umKhwakhwane 107
umKlele 103

umKoto 92
umLahlana 124
umLulama 124
Umncaga 116
umNqabaza 108
umNyenye 116
umQaqongo 98
umQokolo 101
umShungu 113
umSinsana 78
umSintsane 78
Umthele 103
umThumgwa 113
umThungulu 97
umTongwane 113
umTyatyambane 98
umUnyane 92
umValasangweni 107
umVangatane 71
umVusankunzi 97
umXele 103
umXhalagube 40
uNomaweni 68
uNomphumela 107
uNondomela 109
uQongqo 75
Utile 93
uZwathi 83

Vaalbos 122
Vaalbossie 63
Valsboegoe 33
Veerkoppie 58
VERBENACEAE 98, 110
Vlam-van-die-Vlakte 71
Vlieëbos 60
Vuurwiel-speldekussing 82

Weeping Sage 93
White Bauhinia 72
White Bristle Bush 85
Wild Bush Petunia 70
Wild Dagga 81
Wild Fuchsia 109
Wild Pomegranate 75
Wild Rosemary 45, 63
Wild Tulip Tree 123
Wildeappelkoos 101
Wildeboegoe 33
Wildedagga 81
Wildegranaat 75
Wildekamperfoelieboom 124
Wildemalva 56
Wildeparasolblom 110
Wilderoosmaryn 45
Wilderosmarien 63
Wildetulpboom 123
Witkatjiepiering 107
Witolien 95
Witsteekbos 85
Wolharpuisbos 46
Woolly-bearded Protea 90

X **Ruttyruspolia** 'Phyllis van Heerden' 91
Xylotheca kraussiana 125

Yellow Barleria 39
Yellow Pomegranate 117
Yellow Wild Iris 42
Yellowbell Bauhinia 73
Ystervarkbos 117

ZAMIACEAE 104
Zoeloemuishondblaar 62
Zulu Spur Flower 62